MAT LOGIES

DATE DUE

INN OR

A

MATERIALS SCIENCE AND TECHNOLOGIES

Additional books in this series can be found on Nova's website under
the Series tab.

Additional E-books in this series can be found on Nova's website under
the E-books tab.

MATERIALS SCIENCE AND TECHNOLOGIES

INNOVATIVE MATERIALS FOR AUTOMOTIVE INDUSTRY

AKIRA OKADA

Nova Science Publishers, Inc.
New York

NOTICE TO THE READER
The Publisher has taken reasonable care in the preparation of this book, but makes no expressed or implied warranty of any kind and assumes no responsibility for any errors or omissions. No liability is assumed for incidental or consequential damages in connection with or arising out of information contained in this book. The Publisher shall not be liable for any special, consequential, or exemplary damages resulting, in whole or in part, from the readers' use of, or reliance upon, this material. Any parts of this book based on government reports are so indicated and copyright is claimed for those parts to the extent applicable to compilations of such works.

Independent verification should be sought for any data, advice or recommendations contained in this book. In addition, no responsibility is assumed by the publisher for any injury and/or damage to persons or property arising from any methods, products, instructions, ideas or otherwise contained in this publication.

This publication is designed to provide accurate and authoritative information with regard to the subject matter covered herein. It is sold with the clear understanding that the Publisher is not engaged in rendering legal or any other professional services. If legal or any other expert assistance is required, the services of a competent person should be sought. FROM A DECLARATION OF PARTICIPANTS JOINTLY ADOPTED BY A COMMITTEE OF THE AMERICAN BAR ASSOCIATION AND A COMMITTEE OF PUBLISHERS.

Additional color graphics may be available in the e-book version of this book.

LIBRARY OF CONGRESS CATALOGING-IN-PUBLICATION DATA
Okada, Akira, 1960-
Innovative materials for automotive industry / author, Akira Okada.
p. cm.
Includes index.
ISBN 978-1-61668-237-8 (softcover)
1. Automobiles--Materials. I. Title.
TL240.O4325 2010
629.2'32--dc22
2010012201

Published by Nova Science Publishers, Inc. ✚ *New York*

CONTENTS

PREFACE

A wide variety of materials are used in automotive parts, and most of them are mechanical components. Steels are most widely used for the power train and automotive body construction, and the materials and their processing have been steadily improved. The use of aluminum alloys and polymeric materials have been contributed to the reduction in fuel consumption rates due to the lightweight nature. This new book reviews research on material and automotive technologies, as well as how materials work for technological innovations, with particular emphasis placed on the relation between materials science and the performance of automobiles.

Chapter 1

INTRODUCTION

Transportation is an essential function for living creatures, and while members of the human race can walk and run on their feet, the traveling distance is normally short—within several tens of kilometers a day. Riding on the back of gentle and powerful animals such as horses, camels and elephants enabled long traveling distances when carrying a heavy weight. The invention of wheels helped to carry the weight, and horse–drawn vehicles such as wagons and carts became useful for traveling on paved roads, which had hard surfaces. Automobiles employed engine power as an alternative to horses, and the advances in power sources, such as steam engines, electric power, and internal combustion engines, have greatly contributed to the development of automobiles.

TRANSPORTATION SYSTEMS

A variety of transportation systems are now available. Airplanes are advantageous in traveling a long distance, and ships are beneficial for carrying a great amount of cargo with a less expensive cost. Electric trains, such as electric locomotives, trams and metros, are suitable for transportation of a large number of passengers, while the places to visit are limited within the railroad network. On the contrary, automobiles are relatively free to access any places, as long as paved roads and parking spaces are available. This implies that availability of the transportation system depends on infrastructures, which require a great amount of investment for the construction and maintenance, normally served by the government and communities.

An essential requirement for the infrastructure in land transportation systems is to support the loads of vehicles, such as railroads and paved roads, and a great amount of social investment, including the construction of a wide–spreading road network, are required for automobiles to be employed for the major transportation system. On the contrary, the infrastructure investments for airplanes and ships are rather local because of the limited numbers of airports and seaports, respectively. Accordingly, the utilization of automobiles includes charges for the maintenance of paved road surfaces and related facilities, in addition to the initial cost of purchasing the automobile and its running costs, which include fuel consumption and replacement of parts.

POWER SOURCES

Today's major power sources are undoubtedly internal combustion engines and electricity, and the electricity is mostly produced with electric generators driven by steam turbines, gas turbines and hydroelectric generation. Accordingly, the great amount of energy available in the civilized world is derived from heat engines, in which the thermal energy supplied from the combustion of fossil fuels, such as coal, petroleum and natural gas, are transformed into the mechanical and electrical energy. Another source of thermal energy comes from nuclear power, in which the nuclear fission of uranium–235 is used, and high temperature steam produced from the nuclear reactions is used for the rotation of steam turbines to generate electricity. Accordingly, most of mechanical power and the converted electricity, used in the civilized life, are derived from heat engines.

The advances in power sources have greatly influenced the choice for the transportation systems. Airplanes had been powered by gasoline engines since the first flight of the Wright brothers in 1903, and the development of jet engines during the 1940s to 1950s has completely changed the major power source of airplanes from gasoline to jet engines. Efficient generation of electricity enabled the rise of electric locomotives and the decline of steam locomotives. Further advances are expected in the near future to completely change the power sources of automobiles from traditional gasoline engines to advanced ones.

Efforts for improving the efficiency of heat engines may be classified into two categories. One is the addition of small units to engines for recovering the thermal and mechanical energy, and another is to develop the power sources of high efficiency. The former includes thermal energy recovery by means of adiabatic turbo–compound diesel engines and thermoelectric generation from the exhaust thermal energy, and hybrid electric vehicles enabled the recovery of mechanical energy to generate electric power in a course of reducing the vehicle velocity. The latter includes fuel cell vehicles and high temperature gas turbines equipped with regenerators, and the research activities on ceramic gas turbines had been intensive during the 1970s to 1980s. Recent activities may be characterized by a concentration on electric vehicles having advanced batteries, fuel cell vehicles, clean diesel engines and hybrid electric vehicles.

ELECTRONIC CONTROL

Another problem in transportation systems is associated with safety and environmental issues. This is basically derived from the consumption of the huge amount of energy that is required for numerous high-speed vehicles to be driven. The large kinetic energy of a vehicle may cause great impact damage at a traffic accident, and the huge consumption of energy and natural resources may alter the environment in which people live. Electronic technologies have partially solved the problems, and safety devices such as seat belts and airbags actually reduce the impact damage to passengers during a collision. The safety control of a vehicle, such as an antilock braking system, has reduced the possibility of severe traffic damage to passengers, and the electronic engine control enabled the purification of exhaust gases.

Chapter 2

POWER SOURCES

The major power sources of automobiles are gasoline engines and diesel engines, both of which are reciprocating engines, while the differences can be placed on the combustion mechanisms. Advanced engines of superior thermal efficiency have been extensively investigated, including ceramic gas turbines and a variety of thermal energy recovery systems such as adiabatic turbo–compound diesels and thermoelectric generation systems from exhaust gas. However, the development of advanced engines exceeding the performance of gasoline and diesel engines seems considerably difficult at the present time, while electric power has been recently expected to be the next generation of automotive power source.

In this chapter, the principles of power sources are presented. First, the basic concepts of reciprocating engines are described. Second, potential candidates for advanced engines are summarized, including a variety of thermal recovery systems from exhaust and high temperature gas turbines. Finally, the application of electric power to automobiles is presented with a focus on electric vehicles, fuel cell vehicles, and hybrid electric vehicles.

2.1. HEAT ENGINES

The origin of heat engines may be a steam engine that has been developed by T. Newcomen in 1712. Newcomen's engine was used for pumping water in mine workings, and the engine has a reciprocating piston in a cylinder. Steam is introduced in the cylinder to push up the piston, and the vacuum, which is produced with the introduction of cold water into the cylinder to condense the steam, pulls down the piston. The thermal efficiency is, however, very low. This is because the introduced cold water is used not only to condense the vapor but also to cool the cylinder of its large heat capacity. In addition, the thermal energy of hot steam is also consumed for raising the cylinder temperature. In 1769, James Watt greatly improved the thermal efficiency of Newcomen's engine by installing a separate condenser. The steam vapor is condensed in the small room of the separate condenser, and the amount of cooling water is greatly reduced due to the small heat capacity of the separate condenser in comparison with Newcomen's large heat capacity of the cylinder. Watt's engine was applied to blast furnace blowers for steel making.

The steam engines were applied to automobiles, and N. Cugnot produced a steam mobile in 1769. The significant advance in the application of steam engines was, however, not achieved in automobiles but in ships and locomotives. The first ship equipped with a steam engine was built in 1783, and Robert Fulton started a regular passenger service in 1807.

The success in a steam locomotive is associated with a high–pressure steam engine, in which the condenser is eliminated and high–pressure steam is used for alternately pushing the piston to make reciprocating movement in the cylinder. In 1799, R. Trevithick produced a high–pressure steam engine, and a steam locomotive was successfully demonstrated in a limited traveling distance. The failures of cast iron railway and structural parts such as gears were difficult problems to overcome, and the performance of steam locomotives was insufficient at the time. Based on the pioneer work, the first success in steam locomotives was made by G. Stephenson, who opened the first steam locomotive public railway in 1825.

Automotive Engines

E. Lenoir developed a single–cylinder two–stroke gas engine in 1860, and this was equipped with a spark plug for ignition. In 1876, N. A. Otto developed a four–stroke gas engine, which had basically the same structure as today's conventional engines. In 1885, C. Benz and G. Daimler independently developed vehicles equipped with four–stroke gasoline engines. The advantage of vehicles equipped with gasoline engines was powerful over electric cars and steam cars while difficult and dangerous procedures are required for starting engines to rotate crankshafts by hand and technical skills for operating the manual transmissions.

The use of steam–powered cars had been increasing since the early 19th century and was still popular even in the early 20th century, in spite of the troublesome requirements for water supply to generate steam. The advantages of electric vehicles were silent power sources and easy to drive while the short mileage is inevitable. Accordingly, steam cars and electric cars were popular automobiles even in the beginning of the 20th century, and the advantage of vehicles equipped with gasoline engines was powerful over electric cars and steam cars. The performance of gasoline engines has been improved since then, and the gasoline engines became the major automotive power source in the early stage of the 20th century. Further advances in reciprocating engines were conducted by the invention of diesel engines, and the thermal efficiency of the diesel engines is much higher than that of gasoline engines.

The development in gasoline engines is associated with the equipments to assist easy operations. The first useful electric starter was invented by C. F. Kettering and installed in Cadillac in 1912, and easy operation was also due to the advances in transmission technologies.

Developments for Advanced Engines

The Carnot cycle theoretically gives the maximum in the thermal efficiencies of heat engines. As shown in Figure 2.1, the Carnot cycle consists of isothermal expansion, adiabatic expansion, isothermal compression and adiabatic compression. The gas temperature during the isothermal expansion is kept T_1, and the gas temperature is then reduced during the adiabatic expansion to the temperature of T_2. The gas temperature is kept constant during the isothermal expansion process, and the temperature is raised to T_1 after experiencing the adiabatic compression. The thermal efficiency in this cycle is given by

$$\eta = 1 - T_2/T_1 \tag{2.1}$$

It should be noted that the Carnot cycle is an ideal engine cycle and no other engine cycles exceed the thermal efficiency of the Carnot cycle. Theoretical efficiency in other types of internal combustion engines depends on the combustion mechanisms, and the Carnot cycle theoretically gives the highest thermal efficiency.

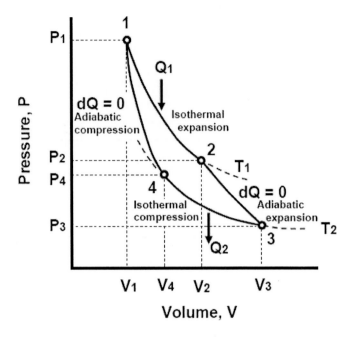

Figure 2.1. Pressure–volume (*P–V*) diagram for Carnot cycle.

Gasoline and diesel engines are thought successful among a variety of power sources that have been developed for automobiles. Nevertheless, automotive engineers have been involved in developing advanced power sources having excellent thermal efficiencies. Gas turbines and adiabatic turbo–compound diesel engines had been considered as a candidate and intensively investigated in the 1970s to 1980s. Electric power sources, such as hybrid electric vehicles, fuel cell vehicles, and electric vehicles, appear as the next generation of candidates, and some of the vehicles have been commercially available since the late 1990s. Along with the development of new power sources, the improvements in traditional gasoline and diesel engines are in progress, and have enhanced the efficiencies of reciprocating engines commercially available.

2.1.1. Gasoline Engines

Gasoline engines have improved greatly in performance and they dominated the early 20th century. Technologies for improving the driving performance were basically derived from mechanical engineering, and these technologies were evaluated in car races in the initial stage of the development. Successful technologies developed in the races trickled down to commercially produced automobiles. Basically, the thermal efficiency of gasoline engines is theoretically expressed by the Otto cycle, and the development in engines was conducted in accordance with theoretical consideration.

Principle of the Otto Cycle

Four–stroke gasoline engines are modeled with the Otto cycle, as shown in Figure 2.2. Let us assume the suction stroke in the Otto cycle. With a start of the suction stroke, the fresh gas is introduced to the cylinder through inlet valves, and the volume reaches to the maximum of V_2 with the ambient pressure of P_2. The inlet valves are closed when the piston is moved to the bottom position at the end of stroke. The gas is then compressed adiabatically to the volume of V_1 corresponding to the compressed position of the piston in the compression stroke, and the pressure is raised to P_3. With a start of the combustion stroke, the gas is suddenly heated with absorbing the heat of Q_1 under a constant volume, and the pressure is raised to the level of P_4. The gas is then adiabatically expanded to the volume of V_2, and the pressure is reduced to P_5. In the final exhaust stroke, the exhaust valves are open to reduce the pressure down to P_2 with releasing the heat of Q_2. The gas in the cylinder is exhausted through the exhaust valves, and the cylinder volume reaches to V_1 under ambient pressure of P_2. The exhaust valves are then closed and the inlet valves are open for starting the suction stroke again.

The theoretical thermal efficiency η in this cycle is given by

$$\eta = 1 - \left(1/r_c\right)^{\gamma-1} \tag{2.2}$$

where γ denotes the specific heat ratio defined by $\gamma = C_p/C_v$ and r_c denotes the compression ratio given by $r_c = V_2/V_1$. Here, C_v and C_p denote specific heats at constant volume and at constant pressure, respectively.

Figure 2.3 shows the theoretical thermal efficiency of the Otto cycles, indicating that the thermal efficiency of the Otto cycle is governed by the specific heat ratio and the compression ratio. Accordingly, high thermal efficiency is achieved under the operating conditions of high compression ratios using lean gas mixtures. This is because the specific heat ratio is high in lean gas mixtures.

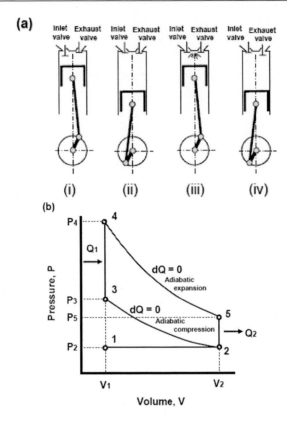

Figure 2.2. Principle of four–stroke gasoline engines.

(a) Diagram of the four–stroke gasoline engine cycle. The gas mixture is introduced to the cylinder through the open inlet valve shown in a suction stroke (i), and the gas mixture is compressed with closing valves in a compression stroke (ii). In a combustion stroke (iii), the compressed gas mixture is ignited with a rise in the pressure. Combustion gas expands its volume and the piston is moved down. After the expansion, the combustion gas is exhausted through exhaust valves in an exhaust stroke (iv).

(b) Pressure–volume diagram ($P–V$) of the Otto cycle corresponding to the four–stroke engine cycle. In the initial suction stroke (Point 1), the gas has the minimum volume of V_1 under the ambient pressure of P_2. The gas mixture is introduced to the cylinder through intake valves, and the volume reaches V_2 under the pressure of P_2 (Point 2). The gas mixture is adiabatically compressed to the volume of V_1 and the pressure reaches P_3 (Point 3). The gas mixture is the ignited for combustion, and the pressure increases to P_4 with maintaining the volume of V_1 (Point 4). The gas adiabatically expands to the volume of V_2 (Point 5). The combustion gas is exhausted through the exhaust valve, and the pressure is reduced to P_2, (Point 2). The cylinder volume reaches V_1 after the end of an exhaust stroke, and the exhaust valve is closed (Point 1).

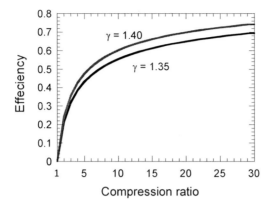

Figure 2.3. Theoretical thermal efficiency of the Otto cycles

Specific Heat Ratio

The values of specific heat of gases are related to the gas constant R in the equation of $C_p = C_v + R$. In addition, the value of C_v is related to the degrees of freedom, and one degree of freedom contributes to $\frac{1}{2}R$ to the specific heat (see Figure 2.4). In monatomic atoms, three translational degrees of freedom are allowed due to simple movements in the three X, Y, and Z–axis dimensions of space. Accordingly, the C_v value of monatomic gases is theoretically $\frac{3}{2}R$.

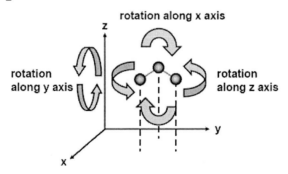

Figure 2.4. Degrees of freedom for gas molecules, translational movements in three directions are allowed for all the molecules while the allowed rotational movements depends on the configuration of molecules. Totally, three degrees of freedom are allowed for monatomic gases and five degrees of freedom are allowed for diatomic molecules.

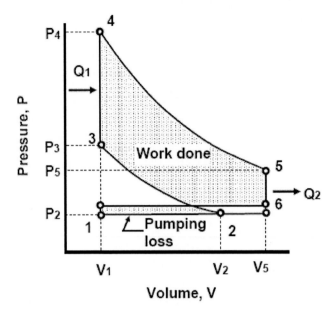

Figure 2.5. Pressure–volume diagram of the four–stroke gasoline engine in the Miller cycle. The compression ratio in the Miller cycle is V_2/V_1, and this is smaller than the expansion ratio of V_5/V_1. The available engine power is also indicated in the diagram, and the actual engine power is reduced by the pumping loss.

The diatomic molecules such as nitrogen and oxygen have five degrees of freedom, leading to the C_v value of $\frac{5}{2} R$. The five degrees of freedom in diatomic molecules comprise three degrees of freedom due to simple translational motion and two rotational degrees of freedom due to geometrical requirements. The specific heat ratio of air is therefore close to 1.4. In the gas molecules of complex structures consisting of three atoms or more, such as H_2O, NH_3 and hydrocarbons, have six degrees of freedom due to three rotational degrees of freedom, and the C_v value is $\frac{6}{2} R$. However, the thermal energy is partially stored in stretching vibration between the bonded atoms at elevated temperatures, and the specific heat becomes large at high temperatures due to the additional vibration modes contributing to the internal energy of gas molecules.

Improving the Efficiency of Gasoline Engines

Thermal efficiency of gasoline engines is theoretically higher at operating in higher compression ratios. However, the operation in the high compression ratios is restricted due to the generation of knocking. The temperature of an air–fuel mixture is raised due to adiabatic compression as well as the boost pressure of turbochargers, and the high–temperature gas mixtures could undergo self–ignition combustion leading to knocking.

In reciprocal engines, high efficiency is achieved in the condition of a high compression ratio with lean burn, and this is principally suitable for the diesel engines. Achieving the conditions in gasoline engines is, however, difficult because the high compression ratio in gasoline engines results in a severe knocking phenomenon and lean gas mixtures may encounter the problem of ignition misfire.

Operating engines in a low compression ratio with a high expansion ratio may be a sophisticated method for improving the efficiency, as shown in Figure 2.5. In this case, the gas volume at the stage of closing inlet valves (V_2 in Figure 2.5) is smaller than that of opening exhaust valves (V_5 in Figure 2.5). Accordingly, the compression ratio is lower than the expansion ratio because the compression ratio is defined by $r_c = V_2/V_1$ and the expansion ratio is $r_E = V_5/V_1$. This situation is realized with the Miller–cycle engines, which require the timing change for closing inlet valves. The valves are closed at slightly different timing from the exact bottom position of the piston. In this case, the compression stage starts from the small gas volume of V_2, in comparison with the maximum volume of V_5 in Figure 2.5.

2.1.2. Diesel Engines

High efficient reciprocal engines were realized with the invention of diesel engines in 1893. In comparison with gasoline engines, the diesel engines are advantageous in large sized engines and have been applied to ships and trucks. The high efficiency of the diesel engines results from the combustion mechanisms, summarized in Table 2.1. The combustion of diesel engines occurs in the conditions of excess air to the fuel, and fuel is injected into the combustion chamber filled with the compressed air.

Table 2.1. Comparison of the gasoline and diesel engines

	Gasoline engines	Diesel engines
Compression stroke	Compression of the gas–fuel mixtures	Compression of the air
Ignition	Ignited with spark plugs	Fuel injection into cylinders
Compression ratio	Low (around 10)	High (around 20)
Fuels	Gasoline of high octane rating to prevent knocking	Diesel fuel of high cetane number for immediate combustion

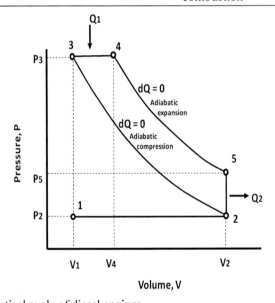

Figure 2.6 Thoretical cycle of diesel engines

Principle of the Diesel Engine

The diesel cycle is the theoretical cycle of diesel engines, and schematically modeled in Figure 2.6. This is similar to the Otto cycle, and the major difference is placed on the combustion mechanism that a compressed gas is to expand with absorbing the heat of Q_1 under a constant pressure. According to the theory, the thermal efficiency is given by

$$\eta = 1 - \left(\frac{1}{r_c} \right)^{\gamma-1} \frac{\beta^\gamma - 1}{\gamma(\beta - 1)} \tag{2.3}$$

where β denotes the cutoff ratio defined by $\beta = V_4/ V_1$.

The theory indicares that the thermal efficiency of diesel engines is governed by the specific heat ratio, the compression ratio, and the cutoff ratio. High thermal efficiency is therefore achieved with the high compression ratio using a lean gas mixture with small heat supply of Q_1 because a low cutoff ratio leads to high efficiency.

It should be noted that the actual engine cycle is close to the partial combination of the Otto and diesel cycles, in which a compressed gas is expanded to the state of higher pressure as well as larger volume.

2.1.3. Thermal Energy Recovery

It is noteworthy that the maximum in the thermal efficiency of the Otto cycle is governed by the knocking phenomenon, which causes the abnormal combustion leading to the damages in engines. Knocking is self–combustion occurring outside the main flame of spark–plug–ignited, and the knocking usually occurs in the conditions of highly compressed gas mixtures. Furthermore, the use of lean gas fuel mixtures may be unsuitable for the Otto cycles because of the difficulties in spark–plug–ignitions to the lean gas mixtures. On the contrary, the combustion in diesel engines normally occurs in lean burn conditions because the combustion is ignited with injecting fuel into the combustion chamber. Accordingly, the diesel engines are able to smoothly operate even in high compression ratios. This leads to higher thermal efficiencies of diesel engines than those of gasoline engines while the diesel engines have disadvantages in difficulty of clean exhaust emission due to the combustion mechanism, and noise and vibration problems result from the high compression ratio.

Factors Influencing the Efficiency of Engines

A basic concept in improving the thermal efficiency of gasoline engines is to operate engines in the efficient conditions of reducing energy losses such as friction losses, leak losses, and heat dissipation through exhaust pipes and cooling water. The combustion of lean air–fuel mixtures is advantageous for improving efficiency of gasoline engines because of the reduction in throttle losses, in especial under low load combustions. Reducing the coefficient of friction in wear parts contributes to saving energy consumption. Leak losses are caused by dissipated fuel resulting from incomplete sealing of piston rings.

However, the most part of energy is dissipated thermally through exhaust pipes and a water jacket, and this is greater than 70% of the total fuel energy.

Intensive work has been done for developing an advanced engine having a superior efficiency to gasoline engines. The thermal efficiency of diesel engines is higher than that of gasoline engines because of no throttle losses, lean burn combustion and high compression ratios. However, in addition to the difficulties in lowering noise and vibration levels as well as purifying exhaust gases, the slightly higher production cost of diesel engines has restricted them to rather limited applications such as trucks, buses and ships. The passenger cars equipped with diesel engines are therefore advantageous in the cases that the total traveling distance is long enough for compensating the expensive production cost.

Adiabatic Turbo–compound Diesel

In turbo–compound systems, turbine rotors are installed in an exhaust gas stream, and the revolutions of the rotors are used for the revolving power of crankshafts. As a result, thermal energy is recovered from high–pressure gases passing through the exhaust pipes. However, the actual applications have been very limited because of the small amount of energy recovered, in comparison with the additional cost for the equipment.

An adiabatic turbo–compound diesel engine was proposed as a candidate for further improvements in diesel engines around the early 1980s. The concept of adiabatic turbo-compound diesel engines is an improvement in turbo–compound engines and is intended to recover the great amount of thermally dissipated energy by means of a turbo–compound system. The basic concept of adiabatic turbo—compound diesel engines is to thermally insulate the cylinder inner walls with ceramic materials having low thermal conductivity and to enhance the thermal energy supplied to exhaust pipes by reducing the thermal energy transferred to cooling water. Intensive work was conducted, and the problems to be solved became clear. One is to develop suitable lubricants that work at elevated temperatures between pistons and cylinders. Another is the low output power resulting from hot cylinders, which resist a great amount of air entering to fill them.

Thermoelectric Generation

There are several candidates to recover thermal energy from exhaust. The turbo–compound systems indeed enable to generate mechanical power from high–pressure gases. Thermal energy in the exhaust can be converted to mechanical energy by additional heat engines such as the Rankine cycle and the Stirling engine. In the Rankine cycle, exhaust temperatures are transmitted to water to produce steam that rotates rotors for generating mechanical power. In the Stirling engine, thermal energy expands the gas filled in the cylinders of the engines to cause reciprocating movements.

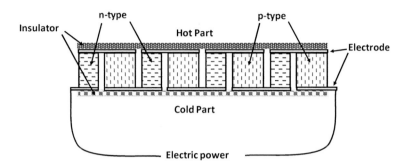

Figure 2.7 Thermoelectric modules

Thermoelectric modules directly transform thermal energy to electrical energy, and work has been conducted to recover the thermal energy dissipated in exhaust gases. The thermal energy dissipated through the thermoelectric elements is partially recovered in a form of electrical energy when thermoelectric modules are placed on the hot surfaces such as the surface of exhaust pipes. The amount of energy recovered is, however, very small because of the poor performance in the thermoelectric modules and the difficulties of designing structures suitable for increasing the amount of heat penetrating through the thermoelectric elements. Figure 2.7 shows thermoelectric modules, in which two types of semiconductors are arranged, and the temperature difference between two ends of the module produces electric power. The generated electric power strongly depends on the performance of the thermoelectric material and the ability of transferring thermal energy from heat sources to the surfaces of the modules.

2.1.4. Gas Turbines

A gas turbine, which consists of a compressor, a turbine rotor, and a combustion chamber, produces a rotary motion in a turbine rotor, as shown in Figure 2.8. The turbine rotor is driven with the hot gas produced in the combustion chamber, and the compressor rotor is linked by sheared axis with the turbine axis of the compressor. As a result, the compressed air is introduced to the combustion chamber, where fuel is supplied to produce the high–pressure combustion gas.

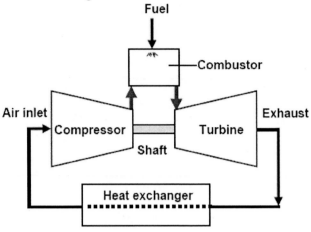

Figure 2.8. Schematic diagram for gas turbines.

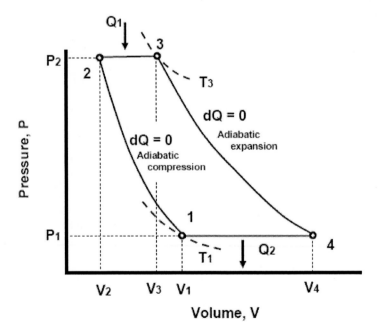

Figure 2.9. Pressure–volume diagram of the Brayton cycle modeled for gas turbines.

The major applications of gas turbines are high efficient electric generators and jet engines for airplanes. In electric generators, the rotation of turbines is directly connected to the electric generator. The jet engines produce the stream of fluid projected into a surrounding medium, and the stream of fluid is produced in the end nozzle placed after the turbine rotors. The gas turbine has a small and simple structure producing huge power, and the rotational power may be applied to automobiles. In addition, gas turbines are potentially of high thermal efficiency in the cases of high gas inlet temperatures. This is the advantage of gas turbines, and the temperature of gas turbines has increased as a result of intensive study on the recent cooling systems for gas turbine components.

Principle of Gas Turbines

The Brayton cycle gives the theoretical thermal efficiency of gas turbines, as indicated in Figure 2.9. Fresh air introduced to the compressor at temperature of T_1 is compressed in an adiabatic manner from the pressure P_1 to P_2, and the volume is reduced from V_1 to V_2. The compressed air of P_2 is then expanded in the combustor under the constant pressure. The combustion produces the heat supply of Q_1, and the combustion gas expands to the volume of V_3 with a raise in gas temperature to the maximum in the cycle, T_3. Hot gas produced in the combustion chamber expands in the turbine rotor, and the pressure is reduced to P_1 with the gas volume of V_4. Finally, the hot gas having heat Q_2 is exhausted from the system. The thermal efficiency of the Brayton cycle is given by

$$\eta = 1 - \left(1/P_r\right)^{\frac{\gamma-1}{\gamma}} \tag{2.4}$$

where P_r is a pressure ratio defined by $P_r = P_2/P_1$.

Accordingly, the thermal efficiency of the Brayton cycle is governed by the specific heat ratio γ and the pressure ratio P_r. The thermal efficiency of gas turbines is enhanced by the high pressure–ratio while the great enhancement is achieved with the use of heat exchangers.

In regenerative Brayton cycles, the heat in the exhaust gas is partially absorbed in the heat exchangers to heat the inlet air since the outlet temperature is much higher than the inlet temperature. As a result, a heat exchanger enhances the absorbed heat Q_1 in the combustion process and reduces the heat Q_2 to the exhaust. Accordingly, the thermal efficiency of the regenerative Brayton cycle is higher than the cycle without heat exchangers, and the thermal efficiency of the Brayton cycle with regenerator is given by

$$\eta = 1 - \left(\frac{T_1}{T_3}\right) P_r^{\frac{\gamma-1}{\gamma}} \tag{2.5}$$

where T_3 and T_1 denote the maximum and minimum temperatures in the cycle, respectively. Note that the high efficiency is realized with high inlet–temperature for the turbine rotor and a low pressure–ratio, while the thermal exchange rate greatly enhances the efficiency in the actual systems.

High Temperature Gas Turbines

The gas turbines equipped with regenerators may have high thermal efficiencies very close to the Carnot cycle when the pressure ratio is closed to unity, and the higher efficiency is achieved with the higher inlet temperature of turbines. Since the combustion in gas turbines continuously occurs, the components used for combustors and turbines are required for excellent durability in the environments of high temperatures. On the contrary, the components used for the combustion chambers and cylinders in reciprocating engines are periodically heated and cooled as a result of intermittent combustions. Accordingly, the components of reciprocating engines in general require less durability at elevated temperatures than those of gas turbines.

The cooling methods of gas turbine components are of key importance for developing high temperature gas turbines. The high temperature components are normally cooled with flowing air penetrating the components, and the air sprouts from the surfaces of components. The component surfaces are also protected by thermal barrier coating, which consists of porous ceramic materials of low thermal conductivity. The molten ZrO_2–Y_2O_3 ceramics are typically sprayed on the surface of Ni–based superalloys. As a result, the inlet temperature of turbines has been greatly raised and recently reaches around 1500°C or higher while the materials used for the high temperature gas turbines, such as the Ni–based superalloys, have limited durability at high temperatures above 1000°C.

Although the high temperature gas turbines have greatly contributed to the high efficiency in power stations, most of electric power is generated by steam turbines, as shown in Figure 2.10. The steam turbines are similar to gas turbines but use water vapor for rotating turbines instead of combustion gases. Thermal energy for steam production is supplied by combustion heat of coal, natural gas and petroleum, and nuclear reactions. The high temperature gas turbines are used in the combined cycle, in which electric power is firstly produced by gas turbines connected to electric generators, and the thermal energy remained in the exhaust gas is used for generating water vapor that is supplied to steam turbines for the secondary electric power generation.

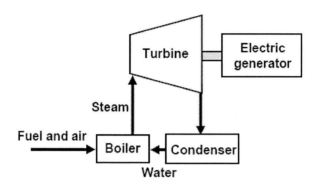

Figure 2.10. Steam turbines for electric generation, in which high temperature steam is produced in the boiler and supplied for rotating the turbine rotors of steam turbines.

Figure 2.11. Automobile installed with a ceramic gas turbine engine that was developed in Sweden and actually tested on public roads in 1982.

Ceramic Gas Turbines

Ceramic gas turbines were intensively considered as a candidate for high–efficiency engines around the 1970s to 1990s. The huge gas turbines have very high thermal efficiency while the small engines exhibit poor efficiency. This is because a part of heat is dissipated from the surface of engines, and heat loss is relatively greater in smaller engines. High thermal efficiency is expected even in small engines when operating at high temperatures.

The concept of ceramic gas turbine was to use high temperature materials for the components of small gas turbines because the inner structure having air paths is difficult to construct for small components. The candidate materials are non–oxide ceramics such as silicon nitride and silicon carbide, which exhibit excellent mechanical properties at elevated temperatures. The brittle behavior has to be overcome, and intensive work has been conducted for developing toughened ceramics and the application of fracture mechanics.

Automobiles equipped with ceramic gas turbines were actually produced, and road tests were successfully conducted. A ceramic gas turbine car that was actually tested on public roads in 1982 is shown in Figure 2.11. However, its performance failed to meet the target. The problems identified in the American project "Advanced Gas Turbine" around 1989 are summarized below.

(1) Failure of ceramic rotors due to foreign object damage
(2) Gas leaking through the joints
(3) Brittle failure due to thermal deformation
(4) Insufficient lubrication of rotating parts
(5) Low strength of ceramic materials at elevated temperatures

A research and development program on 300–kW–class ceramic gas turbines has been conducted in Japan since 1988, and an exceptionally high thermal efficiency of over 42% was demonstrated in 1999 for a turbine inlet temperature of 1350°C. In the project, technical improvements were made for some of the problems.

Elastic supports using silicon nitride springs and ceramic parts joined with ceramic matrix fiber composite materials are shown in Figure 2.12. These technologies were able to minimize the gas leaking at joints with relatively low contact stresses. In addition, silicon nitride having excellent creep resistance was developed by adding Lu_2O_3 as a sintering aid to silicon nitride powder, and this material maintained high strength at the temperatures at which ceramic gas turbines operate.

 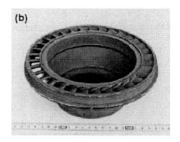

Figure 2.12. Advanced technologies developed in 300kW class CGT project, (a) elastic supports using silicon nitride springs, and (b) joining ceramic parts with ceramic fiber composites (courtesy of Kawasaki Heavy Industries, Ltd.).

2.2. ELECTRIC POWER

The development of electric vehicles was associated with the inventions of electric motors and lead–acid batteries. W. Sturgeon invented an electromagnet in 1825, and M. Faraday discovered the electromagnetic induction effect in 1831. Various types of electric generators and electric motors have been developed since then. In 1859, G. Plante invented the lead–acid battery, and the electric cars practically available appeared in the late 19th century.

Electric powered automobiles are attracting attention in recent years, and a variety of automobiles including electric vehicles, fuel cell vehicles and hybrid electric vehicles, are under development. In electric vehicles, electric power is supplied from power stations to charge batteries, and on–board electric power supply systems are employed for fuel cell vehicles and hybrid electric vehicles.

Electric motors equipped in automobiles are advantageous for the load leveling of reciprocating engines because electric motors operate power sources as well as electric generators used for regenerative brakes. The electric motor produces the power to rotate the wheels during the acceleration of automobiles, and the wheels rotate the motor shaft to generate electric current in the electromagnetic coil of motor in the course of reducing the velocity of automobiles. As a result, the regenerative brake produces electric power that are converted from the kinetic energy of driving automobiles, and the electric power is temporarily charged in rechargeable batteries for driving electric motors.

2.2.1. Electric Vehicles

The lead–acid battery was an invention of great public importance for developing automobile technology and has been used in most of automobiles. However, the recent advances in battery technology have enabled the development of new classes of automobiles, such as hybrid electric vehicles, fuel cell vehicles and electric cars. Some of the vehicles are commercially available while further advances are expected to occur.

The electric cars are very old types of vehicles and actually appeared before the automobiles equipped with gasoline engines. However, the sales have retarded early in the 20th century because of the short mileage and the development of gasoline engines.

The shortage of petroleum supply has lead to the temporary use of electric cars. Actually, numerous electric cars were used in Japan from 1930 to 1950, and 3299 electric cars were in use in 1949. This is approximately 3% of the total number of automobiles at the time in Japan, and this is a very large number since only 42 fuel cell cars are on the roads in FY 2007 in Japan. The large number of electric cars is obviously due to the shortage of petroleum supply during the war in the pacific. Bio–fuel vehicles were alternatively used at the time. The bio–fuel, such as charcoal and wood instead of gasoline, was converted on board to the gas mixture by incomplete combustion, and the vehicles were driven with gas engines. The electric vehicles are actually used even now but very limitedly, such as milk floats in UK.

Rechargeable Batteries

The developments of electric cars have been closely related to the advance in the rechargeable batteries, and a variation of rechargeable batteries are listed in Table 2.2. Nissan Motor Company, for example, developed Electric EV–4 in 1976, which is a hybrid car equipped with a zinc–air battery capable of driving 496 km. March EV–1 was developed in 1983, and the Ni–Fe battery of 370 kg was equipped to drive 160 km. Advanced battery systems, such as Nickel metal–hydride batteries and lithium ion batteries, were developed around 1990. The advances in battery technology lead to the commercialization of hybrid cars, as well as fuel cell vehicles. It should be noted that gasoline engines require small batteries for starting engines and ignitions, hybrid cars and fuel cell vehicles require the batteries of large capacity for load leveling, and electric cars capable of driving long distance require the batteries of huge capacity. The important characteristics for advanced rechargeable batteries are (i) high energy density stored in unit volume and unit mass, (ii) high power density, and (iii) durability or long lifetime. Figure 2.13 shows automotive batteries for electric vehicles.

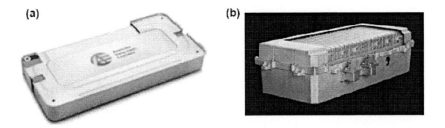

Figure 2.13. Lithium ion batteries developed in Nissan Motor Co. Ltd., (a) laminate type battery, and (b) battery unit for electric cars.

Table 2.2. Rechargeable batteries

	Pb–acid battery	Ni–Cd battery	Ni–MH battery	Ni–Zn battery	Ni–Fe battery	NaS battery	Li ion battery
Anode	Pb	Cd	La–Ni alloys	Zn	Fe	Na	C_6Li
Cathode	PbO_2	NiOOH	NiOOH	NiOOH	NiOOH	S	$LiCoO_2$, $LiMn_2O_4$
Electrolyte	H_2SO_4	KOH	KOH	KOH	KOH	β–alu-mina	Carbo-nate ester

2.2.2. Fuel Cell Vehicles

Fuel cell systems are advantageous in efficiency over heat engines because the maximum efficiency in heat engines is given by the Carnot cycle whilst the maximum efficiency of fuel cells is theoretically given by thermodynamics. The Gibbs free energy change in the reaction can be converted to electric energy while the total energy change during the reaction corresponds to the enthalpy change. Accordingly, the electrical energy produced at room temperature reaction is theoretically 83 % of the total energy, in the case that hydrogen fuel is oxidized by oxygen. This indicates that the potential efficiency of fuel cells is considerably higher than that of heat engines. Table 2.3 lists the several types of fuel cells ever developed, and included are phosphoric acid fuel cells (PAFC), polymer electrolyte fuel cells

(PEFC), molten carbonate fuel cells (MCFC), alkaline fuel cells (AFC) and solid oxide fuel cells (SOFC).

Table 2.3. Fuel cells

	SOFC	MCFC	PAFC	PEFC	AFC
Electrolyte	Y_2O_3–ZrO_2	(Li, Na, K)$_2CO_3$	H_3PO_4	Polymer electrolyte	KOH
Conducting ion	O^{2-}	CO_3^{2-}	H^+	H^+	OH^-
Specific electrical resistance	~1Ωcm	~1Ωcm	~1Ωcm	~200Ωcm	~1Ωcm
Temperature	~1000°C	600 ~700°C	170~200°C	80~100°C	170~200°C
Catalysis	–	–	Pt	Pt	Ni–Ag
Fuel	H_2, CO	H_2, CO	H_2	Pure H_2	Pure H_2

Polymer Electrolyte Fuel Cells

The basic concept using fuel cells for automobiles is to realize the excellent power source of high efficiency, as well as clean exhaust gases. The requirements for automotive power sources are (i) high efficiency, (ii) small power sources, (iii) capable of quick start, and (iv) less expensive.

The phosphoric acid fuel cells are the first commercialized fuel cells and used for small scaled electric power generation including cogeneration applications. However, the problem in application to the power sources of automobiles is the high working temperature around 200°C, the fuel cells have to be heated up to the temperature for starting cells when the phosphoric acid fuel cells are employed for the power source of automobiles. Low working temperatures are preferred for automotive fuel cells, and polymer electrolyte fuel cells, using proton exchange membrane as an electrolyte, meet the requirements. The working temperature is considerably reduced, and the compact fuel cells were produced. The fuel cell vehicles have been commercialized since 2002 in limited uses. The high production cost is the biggest problem, and this is due to the platinum catalyst required for enhancing the reaction at relatively low temperatures. Figure 2.14 shows the internal structure of fuel cell vehicles, the devises such as an inverter, a motor, a fuel cell, rechargeable batteries and a fuel storage vessel for high–pressure hydrogen gas are installed.

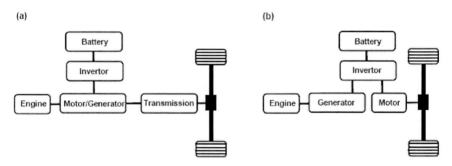

Li-ion batteries

Inverter

Motor   Fuel Cell   Hydrogen storage vessel

Figure 2.14. Structure and components of a fuel cell vehicle, indicating an inverter, a motor, a fuel cell, rechargeable batteries and a fuel storage vessel.

Figure 2.15. Two hybrid vehicle systems, (a) parallel hybrid and (b) series hybrid. The engine in the parallel hybrid system generates mechanical power while the engine in the series hybrid system operates as an electric generator.

2.2.3. Hybrid Electric Vehicle

Hybrid vehicles have double power sources or more, and the hybrid vehicle is typically equipped with a reciprocating engine and one or more electric motors. The hybrid electric cars in recent years employ gasoline engines for main power sources with assistance of electric motors, as shown in

Figure 2.15. The engines generate mechanical power and also charge batteries during driving. In the cases of rapid acceleration and climbing a slope, the large power is required and the combination of engines and motors produces large power. The small power for driving with low speed may be supplied by electric motors, and the batteries are charged during the speed reduction while driving. Load leveling of hybrid electric vehicles contributes to the reduction in fuel consumption rates.

Chapter 3

MATERIALS FOR AUTOMOTIVE STRUCTURES

A wide variety of materials, such as steels, aluminum alloys, glass, ceramics, rubbers, polymers and composites, are used for automotive structure while most of parts are made of steels. Steels are used for major load bearing components in power trains, body structures, and the panels covering automotive bodies. Lightweight materials, such as aluminum alloys and polymer composites, are advantageous in saving fuel consumption due to the weight reduction in vehicles. High temperature materials are required for the parts where hot exhaust gas passes through. Materials used for valve systems and gears require the hard surfaces having excellent wear resistances.

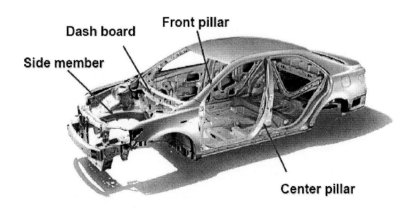

Figure 3.1. Illustrated automotive parts in white bodies.

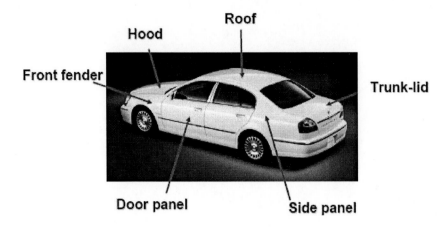

Figure 3.2. Illustrated automotive outer body panels.

3.1. STEELS AND THE RELATED MATERIALS

Steels are commonly used for constructing various automotive structures, and the requirements for the material properties vary in the individual parts. Figures 3.1–3.3 illustrate the representative automotive parts in a white body, an outer body panel structure and an engine, respectively. Load bearing parts such as connecting rods and crankshafts essentially require high strength, and excellent formability is required for producing automotive panels such as hoods, doors, trunk–lids and fenders. The textures of materials are modified to meet the requirements: pure metals are easy to deform in principal, and the composite structures dispersed with precipitates and hard particles are strong.

3.1.1. Texture of Carbon Steels

Carbon steels are capable of producing materials with adjusting mechanical properties by choosing suitable carbon contents and heat treatments. In general, carbon steels of low carbon contents are suitable for constructing the panels with press forming thin sheet materials because pure metals are easily deformed. The strong carbon steels are realized in the microstructure of dispersed fine carbide particles of cementite in a matrix of pure iron, and the distribution and dimension of the carbide phase govern the mechanical properties.

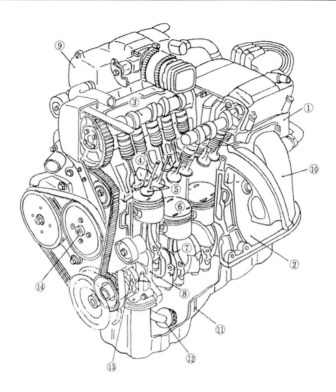

Figure 3.3. Illustrated automotive parts in engines. (1) Cylinder head, (2) cylinder block, (3) camshaft, (4) inlet valve, (5) exhaust valve, (6) piston, (7) connecting rod, (8) crankshaft, (9) inlet manifold, (10) exhaust manifold, (11) oil pan, (12) oil strainer, (13) oil pump, (14) water pump.

Phase Diagram

Figure 3.4 shows the phase diagram of Fe–C. Stable crystal structure of pure iron depends on temperature, and the stable crystallographic form of iron at room temperature is α–iron or a ferrite phase, in which the soluble carbon content is very low. Carbon is present in a cementite phase, which has the chemical composition of Fe_3C. Therefore, carbon steels basically consist of ferrite and cementite phases, and the mechanical properties are governed by the phase ratio and the distribution. The crystal structure of pure iron at elevated temperatures is γ–iron or austenite, in which considerable amount of carbon is incorporated as solid solution.

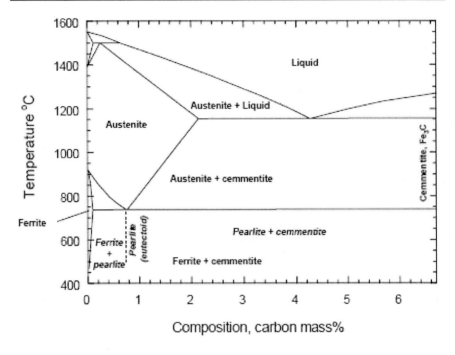

Figure 3.4. Phase diagram of Fe–C system.

In normal cooling conditions, the texture of carbon steels, in which the carbon content is lower than 0.8%, comprises of pearlite and ferrite. The pearlite consists of alternating layers of ferrite and cementite phases and is produced from the eutectoid composition of austenite containing approximately 0.8% carbon when it is slowly cooled below 727°C. Ferrite is produced in carbon steel with very low carbon content, and the material consisting of ferrite exhibits excellent formability.

Heat Treatments

Fine textures in steels are achieved by the heat treatments such as quenching and tempering. The quenching is a process in which materials in the state of austenite phase are rapidly cooled in water or oil. The high temperature phases are kept even in ambient temperature when the high cooling rate is applied. Austenite phase is basically maintained after quenching while austenite phase is readily transformed to the metastable martensite phase below the martensite start temperature. The transformation involves no long–range diffusion process because the martensite phase is a stabilized structure of austenite similar to the crystal structure of austenite.

Fine carbide particles in a matrix of ferrite phase are produced by the decomposition of metastable austenite and martensite phases. Tempered martensite is produced with maintaining the quenched steels typically around 600°C, and the martensite is decomposed into fine ferrite and cementite phases. Similar fine texture of bainite is produced in the medium cooling rate from austenite, and the rate is lower than quenching to produce martensite and higher than gradual cooling to produce pearlite.

3.1.2. Processing for Carbon Steels

A variety of processing methods are employed for metal forming such as casting, forging, and machining. In casting, molten metal is poured into a mold, which contains a hollow cavity of the desired shape. Forging is to shape metal by using localized compressive forces. Hot forging is done at elevated temperatures. The high temperatures make the material to shape easily and less likely to fracture although the surface oxidation is a problem. Producing the smooth surfaces without oxidization is made by cold forging while excellent formability is required for the materials. Machining is a method of partially removing materials to achieve the desired geometry, and power–driven machine tools, such as lathes, milling machines, and drill presses, are used with a sharp cutting tool to mechanically cut the material. In addition, powder ferrous alloys are produced via sintering, and this is advantageous in porous material production and less expensive in production cost due to near net shaping. The former characteristic is applied to the production of sintered oil–retaining bearings, which have been used for various automotive parts in engines, transmissions, and power steering. Sintering is a phenomenon occurring at elevated temperatures, and the densification of porous powder compacts occurs via diffusion processes.

Figure 3.5 summarizes the processing technologies for steels in a diagram of the temperature and applied stress. Metals can be shaped with casting with low or no stresses, and hot forging requires temperatures lower than casting. Cold forging and press forming are conducted in lower temperature but higher stress is required than hot forging. Sintering process is done under low or no applied stresses while it takes a long time for densification.

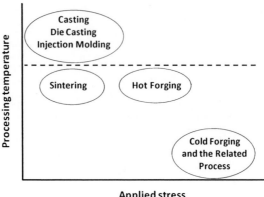

Figure 3.5. Processing technologies for steels in a diagram of the temperature and applied stress. In the diagram, "cold forging" includes various plastic forming process such as press working, extrusion, wire drawing, deep drawing, and form rolling.

3.1.3. Material Selection for Automotive Structures

Carbon steels are widely used for structural materials in automotive structures, and the advantages are less expensive and capability of modifying the mechanical properties by means of heat treatments and the chemical composition control. Hardened steels are extensively used for the load bearing components in engines and transmissions. High strength is achieved by heat treatments to form the fine textures of precipitated carbides in a ferrite matrix, and the hardened surfaces are produced with surface–hardening treatment such as carburizing, nitriding, induction hardening, and shot peening. High strength steels are used for various shafts, bolts, springs, gears and pulleys in engines and transmissions, and also used for tire cord of rubber tire.

Formability and Strength
The basic requirements for sheet steels are formability and strength. The sheet steels for automotive panels require excellent formability, and the application to structural components requires the consistency with high strength. Excellent formability is achieved in the very low carbon steels, in which the strength may be enhanced after the forming process, such as work hardening and precipitation hardening.

Although the texture of pearlite dispersion in a matrix of ferrite phase exhibits high strength consistent with formability, further improvements were made in composite structures of dispersed martensite or austenite in a matrix of ferrite phase. A dual–phase steel has a texture of martensite particles dispersed in a matrix of ferrite phase, and the texture of TRIP steels consists of austenite dispersed in a matrix of ferrite phase (See 3.3 for details). Sheet steels with extremely high strength has basically poor formability, and the high temperature press is usually adopted for the forming process.

Advanced Materials

Limitations in using steels are associated with the severe environments in use. The use in very high temperatures requires refractory materials such as Ni–based supperalloys and ceramics, and the use in chemically hazard environments requires chemically stable materials such as stainless steels, aluminum alloys, polymeric materials and ceramics. The application to wear parts requires hard surface treatments and coating such as carbonizing and hard chrome plating. Hard materials such as cemented carbides and ceramics are also used for wear parts. Replacement by lightweight materials, represented by aluminum alloys, magnesium alloys, titanium alloys, polymeric materials, composites and ceramics, is advantageous for weight reduction of automotive structures, leading to the improvement in the fuel consumption rates.

3.2. STRUCTURAL MATERIALS
FOR DYNAMIC COMPONENTS

The movements of pistons in cylinders are transferred to crankshafts through connecting rods and piston pins, and the rotating powers of crankshafts are finally transferred to the driving axles through transmissions, which consist of an assembly of gears and associated parts. In these mechanical systems, structural materials are required to endure the stresses.

3.2.1. Steels for Engines

The components such as crankshafts and connecting rods are typically produced with hot forging of steel bars at 1200°C, and high strength is endowed by the heat treatment to form the fine dispersed carbide phase in a matrix of ferrite. Martensite phase, which is formed as a metastable phase in a rapid cooling process from austenite phase, contributes to hard and strong properties of the steels, and the brittle aspect in the martensite is modified during the tempering process, where the martensite is decomposed and toughened with slight degradation in hardness. Note that the martensite formation is restrained in the inner parts of large components due to the low actual cooling rates. The transformation to martensite may be enhanced in low alloy steels, which are based on carbon steel and contain several percents of alloying elements such as Cr, Ni, and Mo. The low alloy steels are chosen for manufacturing relatively large components that require high strength, in the cases that the properties of less expensive carbon steels are not enough for the parts.

Crankshafts

Improvements in recent automotive structural steels seem to focus on enhancing fatigue strength and wear resistance with reducing the production cost. This includes the reduction in the amount of expensive elements contained in the materials and the use of less expensive treatments for strengthening. The materials for crankshafts, in which high fatigue strength is in particular required, have been modified to meet the requirements.

Fatigue is mechanical degradation occurring in a material subjected to cyclic loading. In metallic materials, small cracks are generated during the cyclic loading, and the coalescence and propagation of cracks lead to the final failure. Fatigue resistance basically depends on materials, and the surface conditions of materials greatly affect the resistance because the fatigue crack starts from the surface.

Hot forging is usually chosen for the production of high strength crankshafts, and the machined surfaces are hardened to enhance the fatigue strength. A piece of steel, which is heated to the temperature around 1000°C, is placed in a die, and a hammer is dropped on the steel to fill the die cavities. The surfaces are then machined to have accurate dimensions with a removal of the oxidized surface layers. The surfaces of crankshafts are then hardened, and candidate methods for surface hardening of steel typically include (i) carburizing, (ii) high frequency hardening and (iii) soft nitriding.

Carburizing is a process to enrich the carbon content of the surface layer and usually applied to low carbon steels. The materials are quenched to produce hard surface layers after carbon diffusion to the steel surface. High–frequency hardening is an operation to heat the surface of steel components by induction heating, and the surface layers are hardened with maintaining the internal structures. The gas soft nitriding process involves diffusion of both nitrogen and carbon to the surface of steels, and this is less expensive due to no requirements for tempering and the low process temperature of 600°C, in comparison with carburizing treatment.

The surfaces of crankshafts are typically hardened in combination of several techniques including frequency hardening, surface rolling, and the gas soft nitriding.

Connecting Rods

Hot forging and ferrous powder metallurgy are representative methods for producing the connecting rods. High strength low alloy (HSLA) steels are recently applied to the production of connecting rods due to the advantage in cost reduction. This is because the HSLA steel is toughened with precipitation and advantageous in achieving high strength requiring no heat treatment processes, and the details are described in 3.3.2.

3.2.2. Steels for Transmissions

The engine power generated in the combustion of fuel is transferred to the rotation of crankshafts, and the power is partially used for operating auxiliary machines. High rotational speed of crankshafts is reduced in transmission systems, which are used for increasing torque with reducing the rotational speed to rotate the automotive wheels. The rotational reduction is usually made by the combination of spur gears, and the surface hardening of the spur gears is achieved with the vacuum carburizing process, followed by shot peening.

Manual and Semi-automatic Transmission Systems

A manual transmission requires the complicated procedure such as clutch operation and gearshifts. The rotation of the main drive shaft, which is connected to the flywheel at the end of the crankshaft through the crutch, is finally transferred to the rotation of wheels with reducing the rotational speed. A manual transmission system comprises of complex gear mechanisms and a mechanical clutch.

Semi–automatic transmission systems have no requirements for complicated procedures of clutch operation because of no clutch pedals. A variety of automated clutch operation systems have been developed while the transmission mechanism is basically the same as the manual transmission, and the dual clutch transmission system has a twin clutch gearbox instead of the torque converter.

Automatic Transmission Systems

Automatic transmission system is beneficial for the drivers due to no requirements for gearshift operations. The system typically consists of a torque converter and planetary gears, and the fluid in the torque converter produces the connecting force between the engine and the transmission. The force becomes large with an increase in the circulating speed of fluid in the torque converter. The planetary gears enable to change the combination of coupling gears connected to the input and output shafts, and a pair of combination are chosen from three gears of a sun gear, a planet carrier and a ring gear. In the planetary gear system, the rotation of the input shaft is able to connect either shaft chosen for the output shaft by stopping the rotation of the unused shaft.

Gearshift in the automatic transmission system is controlled with hydraulic pressure, which is governed by an electronic control unit with a help of electromagnetic valves. Note that the hydraulic pressure is raised with a pump driven by an electric motor for pressure control employed in the system, and the pressure is transferred with operating the electromagnetic valves controlled by the electronic control unit.

Continuous Transmission Systems

The large numbers of gear sets have to be prepared for realizing a smooth transmission. Instead of using infinitive numbers of gear sets, continuous transmission systems have an advantage in varying the gear ratio continuously. The belt combines two rotating pulleys having different radii, which are continuously changed in the belt continuous variable transmission systems.

The belt consists of several hundreds of steel elements, made of maraging steel sheets that are a special class of low carbon ultra–high strength steels based on the composition of Fe–18%Ni because extremely high strength is required for belt materials. The precipitation of inter–metallic compounds such as $Ni_3(Ti, Al)$ and Ni_3Mo contributes to the ultra high strength.

Carburizing treatment is typically applied to the surface hardening for gears and pulleys. The materials are heated around 950°C in carbon containing gases to absorb carbon element in the surface layer of 0.5 to 1 mm thickness, and subsequently quenched. Resultantly, the high carbon surface is transformed to hard martensite while the inner part remains ductile responsible for excellent impact resistance due to low carbon contents.

3.2.3. Valve Trains and Wear-Resistant Materials

The valve operation is controlled with the valve trains, and a variety of valve train systems have been developed. Valve systems enabling smooth infiltrating and exhausting of gases in combustion chambers are important for reciprocal engines operating at high revolution speeds, and the systems have been mechanically improved. The side valve systems used in the early ages of the automotive industry were turned into more efficient systems, and double overhead camshaft systems with four valves per cylinder have been widely used in recent passenger vehicles. This is because the overhead camshafts allow valves to open and close smoothly, and the multi–valve system is beneficial for exchanging gases in the combustion chambers because of the large opening areas. Valve weight reduction is also effective for smooth opening and closing, and lightweight materials, such as silicon nitride and titanium–aluminum alloys, have been investigated as possible replacements for the heat resistant steel currently used and Ni–based superalloys for high performance exhaust valves. The composition of the heat resistant steel is based on the Fe–Cr alloy containing Ni, and the heat resistance is enhanced with the alloying elements of Ni, Cr and Co. Gamma titanium aluminide valves are used in commercial vehicles, and silicon nitride valves have been used in very limited numbers in formula racing cars.

Wear-Resistant Materials

Hard materials are adequate for the parts in valve systems that experience wear, and surface hardened steels have been mostly used. Alloy steels containing Cr and Mo are typically used for valve lifters, and the surfaces are hardened with coated layers of TiN and CrN.

Ceramics are harder than metals, and silicon nitrides have been used for the valve system since the 1980s. Maintenance–free operation enabled by excellent wear resistance is an advantage of ceramic parts. In vehicles that travel long distances, such as taxicabs and trucks, worn parts may need to be replaced periodically. Ceramics can eliminate such troublesome replacement procedures. Some tappets used for truck diesel engines are shown in Figure 3.6. The use of ceramics for wear parts has enabled long replacement–free durability due to their excellent wear resistance.

Materials of Low Friction

Lowering the coefficient of friction reduces the energy consumption dissipated by friction. Accordingly, recent requirements for wear parts have been not only wear resistance but also low friction. Surface coating with TiN and CrN was intended to lower the coefficient of friction. Hydrogen–free diamond–like carbon (DLC) coating on valve lifters is recently used for reducing the coefficient of friction in normal lubricant oil. Figure 3.7 shows a DLC–coated valve lifter and the coefficient of friction as a function of hydrogen content in the DLC layer. It is clear that the coefficient of friction becomes considerably low as the hydrogen content in the DLC layers decreases, and very low friction is achieved with hydrogen–free DLC–coated materials produced by arc–ion plating.

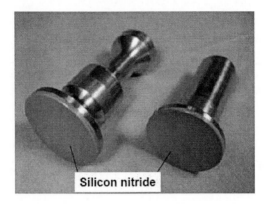

Figure 3.6. Ceramic tappets used for diesel engines (courtesy of NGK Spark Plug Co., Ltd.).

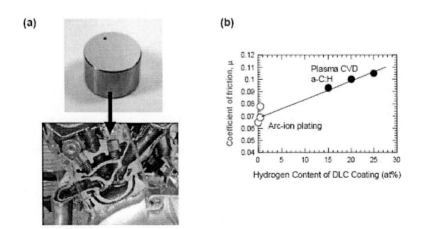

Figure 3.7. DLC–coated valve lifter: (a) a view and (b) the coefficient of friction as a function of hydrogen content in the DLC layer.

3.2.4. High Temperature Materials for Engines

In reciprocating engines, combustion chambers are suffered with cyclic heating by the combustion of engines while the average temperature of the combustion chambers are maintained relatively low due to the cool gas mixture introduced in a next stroke. The hot gas produced in the engines passes through the particular parts, including the exhaust valves, the exhaust ports, the exhaust manifolds, the exhaust pipes and the turbocharger rotors. Although the gas temperature is gradually reduced by dissipation of thermal energy, heat resistant materials, such as cast iron, heat resistant steels and superalloys, are used for the hot parts.

With an increase in the exhaust gas temperatures as a result of improvements in thermal efficiency of automotive engines, the requirements for materials are becoming severe. Materials for exhaust manifolds are changing from gray cast iron to Cr alloy steels, and advanced materials, such as ceramics and Ti alloys, have been applied to high temperature dynamic parts.

Turbochargers

Supercharger systems enable the generation of extraordinarily high engine powers due to the capability of producing high–pressure air in the engine cylinders. Turbochargers, which are turbine–driven forced induction superchargers powered by exhaust gases, use a turbine rotor driven by the gases from the engine exhaust manifolds, and an impeller linked by a shared axle with the turbine compresses ambient air to deliver it to the engine's air intake manifold.

Turbocharger systems are advantageous for yielding extraordinarily high powers through the addition of small turbo–units to engines. However, there is an inevitable delay between the intention to accelerate as expressed by stepping on the accelerator pedal and the actual acceleration of the automobile. This turbo–lag is caused by the time required for the turbine to reach the speed required to supply boost pressure. Reducing the rotor's inertial mass is an effective way to shorten turbo–lag. It is noteworthy that supercharger systems are very suitable for diesel engines, while application to gasoline engines may cause problems resulting from the knocking phenomenon in spite of the great advantage in high output power. In general, gasoline engines require fuel specified with a high anti–knocking index because high efficiency is achieved when the engines are operated at high compression ratios.

The requirements for turbine rotors of turbocharger systems are to endure at elevated temperatures and lightweight because the rotation speeds of the rotors are very high and driven by high temperature exhaust gas. Lightweight is also required for compression rotors, and cast aluminum alloys are commonly used.

Ni–based Superalloys

Carbon steels are unsuitable for the use in elevated temperatures because of the toughned mechanism resulting from a composite structure of cementite dispersion in a matrix of ferrite phase, which is realised at the temperature below 727°C. The improvements were made by alloying the elements of high temperature resistance, such as chrome and nickel to steels.

The heat resistant steels typically contain several percents of chromium and small amount of other metallic elements such as nickel, cobalt and tungsten. To improve the durability at high temperatures, such as applications to gas turbine blades, superalloys are employed because of excellent mechanical strength, creep resistance and oxidation resistance at high temperatures, and nickel and cobalt are the common base alloying elements for superalloys. In automobiles, Ni–based superalloys have been used for the parts of high temperatures such as turbocharger rotors.

Ceramics

The advantage of ceramics is lightweight in comparison to superalloys while the brittle behavior has limited the application to structural components at high stresses. The fracture mechanics and the processing technology of high strength ceramics are progressed in the 1970s to 1980s, and enabled the use of ceramics as structural components. Figure 3.8 shows a ceramic turbocharger made of silicon nitride. The advantage of using ceramics for turbocharger rotors is to shorten the turbo–lag because silicon nitride is lighter than the traditional Ni–based superalloys. Figure 3.9 shows the comparison of revolution speeds between ceramic and metal rotors, and time taken to reach 10,000rpm is 36% shorter for the ceramic rotor.

Ti and the Inter-metallic Alloys

Alloys, such as magnesium, aluminum and titanium, are basically advantageous in lightweight while high temperature properties are insufficient. Among them, titanium alloys are used for exhaust valves and mufflers, although limited applications are obviously due to the high material cost in comparison to steels. A two–phase alloy of Ti–6Al–4V has excellent mechanical properties and used for lightweight automotive components of engines such as inlet valves and connecting rods. Gamma titanium aluminide TiAl is an ordered inter–metallic alloy, which is characterized with a low density of about 4.0 g/cm^3, excellent mechanical properties and oxidation resistance at elevated temperatures. Gamma titanium aluminide has been applied to the turbocharger rotors.

Figure 3.8. Ceramic turbocharger made of silicon nitride.

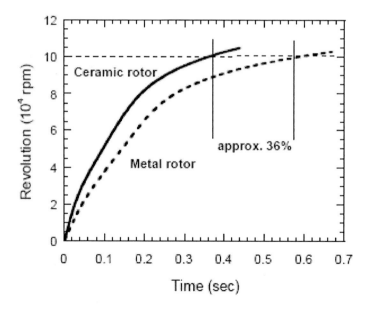

Figure 3.9. Comparison of the revolution rates for ceramic and metal turbocharger rotors.

3.2.5. Steel Filaments for Coil Spring

Springs are used for engine valves and suspensions, and high strength is required for the spring materials. Engine valves are pushed with small coil springs to the holes of the cylinders. Suspension is used for reducing vibration transmitted from road roughness. The major components of suspension are spring and shock absorbers, and large coil springs are used for passenger cars.

Small coil springs are usually made of hardened steel. The traditional piano filaments have been replaced by oil–tempered steels, and cold drawn stainless steel wire is recently used. Valve springs are typically produced with coiling the oil tempered steel filaments of 2mm in diameter.

Large springs such as coil springs for suspension are fabricated in high temperature processes and heat treatment for hardening after the fabrication. The steel filaments are heated at 860 to 900°C and plastically formed into the coil shape with subsequent quenching into oil. The coil is then tempered for modifying mechanical properties.

3.2.6. Aluminum Alloys for Engines

Lightweight materials contribute not only to reducing fuel consumption rates but also enhancing driving performance. Lightweight characteristics are especially useful for the parts that move at high speeds because the reduction in inertial mass enables the mass reduction of supporting parts. Mass reduction is thus effective in the moving parts such as intake and exhaust valves, pistons, piston pins, connecting rods and turbocharger rotors, and intensive work has been performed.

Densities of aluminum and magnesium alloys are considerably lower than those of steels, and aluminum alloys have been successfully applied to pistons, cylinder blocks, cylinder heads, cylinder head covers, oil pans, intake manifolds, wheel covers, hoods and suspension arms. It is noted that some of aluminum alloy parts are replaced with polymeric materials, and the polymeric materials have successfully replaced the cylinder head covers and intake manifolds, and the details are presented in 3.4.2.

Pistons

Aluminum alloys for pistons are required to have a low thermal expansion coefficient to minimize the clearance change between the piston and the cylinder during heat cycles. This is because the dimensional changes in water–cooled cylinders are relatively small during heat cycles in comparison to those of pistons. Accordingly, the gap in cold engines is large in comparison with that in the hot conditions. The addition of silicon is effective for reducing thermal expansion coefficients, and Al–alloy containing 12% Si (Al–12Si), which is close to the eutectic composition, is typically used for pistons of four–stroke engines. Note that the linear thermal coefficient of the Al–12Si alloy is 19×10^{-6}/K, which is considerably lower than 23.5×10^{-6}/ K of pure aluminum.

Cylinder Blocks

Cylinder blocks and cylinder heads for gasoline engines are commonly produced with die casting aluminum alloys because of lightweight. The cylinder blocks normally have cylinder liners of cast gray iron with a thickness of 1.5 to 3.5 mm.

3.3. PANELS AND AUTOMOTIVE STRUCTURES

The steels of large elongation ability have low strength, and high strength steels exhibit poor elongation. Figure 3.10 shows the total elongation and ultimate strength for a variety of sheet steels. Interstitial–free (IF) steels are characterized to exhibit large elongation and low strength among them, and tempered martensite steels have very high strength with small elongation. Note that advanced steels such as dual phase (DP) steels and transformation induced plasticity (TRIP) steels exhibit relatively large elongations maintaining high ultimate strengths.

The major requirement for automotive outer and inner panels is the compatibility of formability and strength. This requirement may be satisfied with materials having the yield stress lower than 250MPa, and the value corresponds to the tensile strength of 340 MPa. These requirements are basically realized in the steel materials, in which the phosphorus element is added to IF steels for solution hardening. Bake–hardening (BH) steels are used for the panels of doors and hoods, and DP steels are applied to wheel discs and strength members. The hot rolled DP steels with the high tensile strength of 780 MPa have been applied to automotive chassis, and the cold rolled DP steels with the very high tensile strengths of 980 MPa have been used for door beams and bumper beams.

Surface Coating

Corrosion resistance is important for the underbody structures, and two classes of zinc coated sheet steels have been developed. One is coated with electrolytic deposition, and another is produced with dipping in zinc melt followed by annealing to produce Fe–Zn layers at the interface.

Paint is coated on an automotive body, and the automotive paint in general consists of three layers. The lower layer is produced with cationic electro–deposition, and the second and third layers are produced with electrostatic spray painting using spray guns. The paint is then baked for hardening.

3.3.1. Formability of Sheet Steels

The excellent formability is achieved in very low carbon steels, and the excess carbon is reduced to the level of 10 ppm by applying the vacuum degassing process to the steel melt produced in the converter process. In the converter process, the chemical reaction occurs between oxygen gas and the excess carbon, and gaseous carbon monoxide is eliminated from the melt while the small amount of carbon is still present in the melt.

Interstitial-free (IF) Steels

The carbon contents in IF steels are extremely low, and the residual carbon is chemically combined with the elements having a strong affinity to carbon such as niobium and titanium. As a result, carbon is completely eliminated from the interstitial sites of the steel matrix, forming fine carbide particles dispersed in the steel matrix. The modified microstructure consisting of soft matrix with hard precipitates has excellent capability of large deformation in comparison with homogeneous solid solutions.

Bake-hardening (BH) Steels

BH steels are similar to the IF steels but contain an allowable level of carbon in the interstitial sites. The steel sheets have an excellent formability, and are plastically deformed into complex shapes. The yield stress is enhanced during the process of baking enamel paints in the temperature range of 150 to 200°C. In this process, the ability of plastic deformation is reduced because the dislocation movement is restricted, and materials are hardened. Dislocations accumulated during the plastic deformation are fixed in the baking process with reacting to the elements of C and N present as residual impurities in the steel.

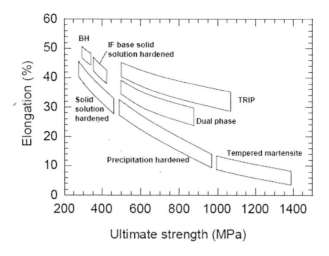

Figure 3.10. Elongation and ultimate strength for a variety of sheet steels.

3.3.2. High Strength Sheet Steels

The first attempt to develop high strength sheet steels for the automotive industry was to employ solid solution hardening, and soluble elements, such as Mn and Si were introduced to carbon steels to develop the high strength steels. The tensile strength achieved was not as high as expected, and the maximum strength was around 400 MPa.

High Strength Low Alloy (HSLA) Steels

The second attempt for developing high strength sheet steels was conducted in HSLA steels, which are also known as microalloyed grade. This class of alloys contains small amounts of vanadium, niobium and titanium to form fine precipitates, and high tensile strength is realized in comparison with that of solid solution alloys, in spite of the very small addition of alloying elements, typically less than 0.1%. Note that the precipitates are indirectly responsible for the high strength, and the major cause for the high tensile strength is not precipitation toughening but the precipitation induced fine–grained microstructure of ferrite phase, which is the major phase in low carbon steels.

Dual Phase (DP) Steels

The third attempt in high strength steels was to develop the microstructure consisting of soft and hard phases. DP steels, consisting of a ferrite matrix and the secondary phase of martensite, exhibit a high work hardening rate resulting in the increased ultimate strength. The ferrite phase contributes to formability, and strengthening is due to the martensite phase. Note that the high initial work hardening rate is typical in materials consisting of a hard dispersion in a soft matrix. The DP steels are produced in the continuous casting, where the molten steel is solidified in the mold, and the martensite phase generates in the subsequent rolling process due to its high cooling rate.

Transformation Induced Plasticity (TRIP) Steels

TRIP steels are further improved ones, which consist of ferrite, bainite and retained austenite phases. The retained austenite enhances the plastic deformation during forming while the high strength is endowed to the steels experiencing the plastic forming process where transformation occurs from austenite to martensite. The texture of TRIP steels is stabilized during the annealing process around 400°C, and the bainite is generated during the process. The formation of ferrite and bainite enriches the carbon content in the retained austenite to the level of 1 %, leading to the stabilization of austenite phase at room temperature. It is noteworthy that the additions of silicon and aluminum to steel are effective for further stabilization of the austenite phase.

Tempered Martensite

High pressure is in general required for the cold press forming of high strength sheet steels though cracks may generate during the process due to the brittle behavior associated with the high strength. Press forming at elevated temperatures enhances the formability of the materials and reduces the pressure required for forming.

Hot stamping is a press forming method for the high–strength steel sheets, and the steel sheets are heated around 900°C and pressed in a cold die with relatively low pressures in comparison with the cold press forming. The formed sheets are rapidly cooled in the die, and the high cooling rate in a hot stamping process leads to martensite transformation resulting in high tensile strength after forming. The oxidized surface generated in the high temperature process is removed in the following process of shot blasting. Hot stamping is applied to the steels with tensile strengths typically higher than 1500 MPa.

3.3.3. Aluminum Alloy Panels

Aluminum alloys are advantageous for constructing automotive panels such as hoods, trunk–lids, and doors because of lightweight, and the requirements for panel materials are formability and strength. Accordingly, the addition of reinforcing elements to aluminum alloys is low for the plastic deformation process and the composition is chosen for enhancing the ability of strengthening the materials during aging around 150°C. Inter metallic compounds, such as Mg_2Si and $CuAl_2$ are precipitated during the process and strengthen the materials. Note that the famous high strength Al alloy is duralumin, which is an Al–Cu alloy used for aircraft body, and fine particles are precipitated during aging process after solid solution treatment.

3.4. POLYMER AND COMPOSITE MATERIALS

Polymeric materials are advantageous in weight reduction and successfully applied to the parts of relatively low stresses, such as bumper covers and instrumental panels. The additions of elastomer and hard particles to the polymeric materials improve the mechanical properties such as impact resistance and strength. Engineering plastics have rigid molecular structure of exhibiting excellent mechanical and thermal properties and are applied to the engine and mechanical parts.

3.4.1. Polypropylene

Polypropylene (PP) is a thermoplastic polymer produced by polymerization of propylene ($H_2C=CHCH_3$), and methyl groups ($-CH_3$) are arranged in polymer chains. Ziegler–Natta catalysts, typically $TiCl_3$–$Al(C_2H_5)_3$, are used for promoting the polymerizing reactions, and the resultant polymer is mostly isotactic polypropylene, in which the methyl groups are consistently on one side of polymer as shown in Figure 3.11. Note that polymer of regularly arranged molecular configuration has excellent mechanical and thermal properties in comparison with randomly arranged one. Polypropylene is the most common polymer for automobiles and is used for the parts such as bumpers and instrumental panels.

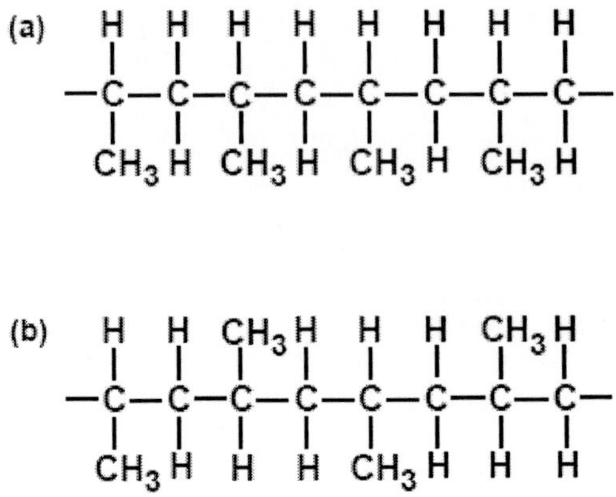

(a)

$$-\overset{\overset{\displaystyle H}{|}}{C}-\overset{\overset{\displaystyle H}{|}}{C}-\overset{\overset{\displaystyle H}{|}}{C}-\overset{\overset{\displaystyle H}{|}}{C}-\overset{\overset{\displaystyle H}{|}}{C}-\overset{\overset{\displaystyle H}{|}}{C}-\overset{\overset{\displaystyle H}{|}}{C}-\overset{\overset{\displaystyle H}{|}}{C}-$$

(b)

Figure 3.11. Regularly arranged molecular configuration of polypropylene, (a) isotactic polypropylene and (b) syndiotactic polypropylene.

Polypropylene Bumpers

A traditional type of passenger car bumpers is made of steel while recent automobiles employ PP composite bumpers. Figure 3.12 shows the structure of the PP composite bumper, which consists of metal bumper beams, polymer foam, and PP composites cover. The PP composite bumpers are advantageous for impact reduction to pedestrians when the traffic accidents occur. The soft materials of polymer foam and PP composites skin cover absorb the impact energy, and the impact damage is reduced. The PP composites, which are used for the outer skin covers of the bumpers, are typically comprised of polypropylene copolymer, elastomer, and talc as a filler material. Improvements have been conducted toward high modulus of polypropylene copolymer, excellent formability in injection molding and the improvement in impact resistance.

The higher elastic modulus of polypropylene was achieved by higher level in the crystallization, owing to the progress in the olefin polymerization catalyst. Ziegler–Natta catalyst has been successfully used for olefin polymerization, and the advanced catalyst of metallocene is advantageous in controlling the configuration of molecular chains. High impact resistance is endowed by the addition of elastomer to propylene–ethylene copolymers, and the polypropylene bumpers have greatly contributed to the weight reduction in automobiles, as well as mitigating the accidental damages to pedestrians.

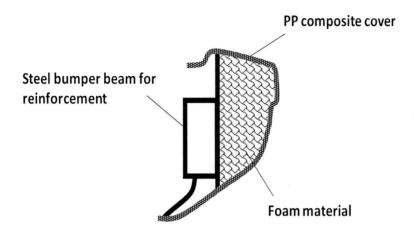

Figure 3.12. A cross section view of PP composite bumpers, consisting of metal bumper beams, polymer form, and PP composites cover.

Instrumental Panels

Polyurethane foam covered with poly–vinyl–chloride skin was typically used for instrumental panels while the recent trend is to use the material adaptable for recycling and cost reduction. As a result, polyolefin materials, such as PP foams with thermoplastic olefin skins, are used in the recent instrumental panels.

3.4.2. Engineering Plastics

The engineering plastics, represented by polyamide (PA) and polyoxymethylene (POM), are used in the parts requiring high performance in thermal and mechanical properties, including intake manifolds, intercooler tanks, cylinder head covers, and bearings.

Polymeric materials used for engine parts require the durability at elevated temperatures, and glass fiber polyamide composites have replaced the aluminum cast alloy parts, which were used for the intake manifolds and cylinder head covers. The advantages in the replacements are not only weight reduction but also noise reduction due to the reduced surface roughness and vibration damping ability of polymeric materials.

3.4.3. FRP Composites

Fiber–reinforced plastic (FRP) composites are formed with embedding continuous fibers in a resin matrix that binds the fibers together. Numerous combinations exist among fibers and resins while commonly used are glass FRP (GFRP) composites and carbon FRP (CFRP) composites.

3.5. AUTOMOTIVE GLASS

The general requirements for the windshield glass are transparency and high strength. Glass is an amorphous solid, and the most common type of glass is soda–lime glass, which is produced from the melts involving sodium carbonate, limestone and silica (silicon dioxide).

3.5.1. Windshield Glass

Tempered glass was the first application of toughened glass to windshields. The tempered glass is produced in a rapidly cooling process with forced air, and the compressive stress generated on the surface of the glass enhances the strength. The problem in the tempered glass applied to windshields is that many separated fragments are produced when the windshield is broken.

Laminated glass is a type of safety glass and consists of two curved sheets of glass with a plastic interlayer of polyvinyl butyral laminated between them. The glass layers thus keep bonded to the interlayer when the windshield is broken. Most of the automotive windshields are of laminated safety glass, and the windshields are glued on the window frame.

3.5.2. Transmittance Control in Glass

An additional requirement for the glass windows is to protect cabins from the strong sunshine without degradation in transparency. The requirement is to reduce the transmittance in the wavelengths of ultraviolet and infrared light with high transmittance of visible light.

Transmittance in glass is controlled through reflection and absorption. Low transmittance of ultraviolet is achieved by the material that has a band gap in an ultraviolet region close to the frequency of the visible light. Glass containing mixed oxides of cerium and titanium absorb ultraviolet effectively while the commercially produced glass absorbing ultraviolet light contains ultraviolet absorbers, such as benzotriazole and benzophenone, in polymer films inserted between plates of laminated glass.

Low transmittance of infrared may be made by selective reflection, and the surface of infrared reflection glass commercially available is coated with thin multi–layers produced by vacuum vapor deposition. Incident light interferes in multi–layered structures, enabling the reflection of electromagnetic wave in infrared frequency. The multi–layered structure enables the rapid change in transmittance in the infrared region, in comparison with semiconductor coating.

In semiconductor materials, the electromagnetic wave penetrates the materials when the frequency is higher than the plasma frequency of the material, and reflection occurs for the electromagnetic waves of lower frequencies. The plasma frequency of semiconductor materials is normally in infrared frequency, and the details are presented in A4.7.

The typical impurities absorbing light in soda lime glass are ferrous oxide (FeO) and ferric oxide (Fe_2O_3). With an increase in the concentration of FeO in relation to Fe_2O_3, the color of the glass shifts from a yellow–green to blue–green. Dark tinted glasses are used for automotive windows and sunroofs, and transition metal oxides are added to the glass for generating dark color.

3.5.3. Water-Repellent Glass

Water–repellent glass was recently developed. Glass can be water–repellent when the surface is coated with an organic water–repellent material. The enhanced durability of the coating was made by water–repellent glass, and the glass is coated with fluorine compounds strongly bonded to the surface of glass.

3.6. TIRE AND BRAKES

Tires in automobiles are directly contacted to the road surface, and large frictional force is required for reducing speed. The rotation of tire wheels is controlled with braking systems. The disc brake is typically used in passenger cars, and brake pads are forced against both sides of the disc to stop the vehicle. The drum brake is another type of brake, and brake shoes are press against the inner surface of a rotating drum connected to a rotating wheel.

3.6.1. Tire

Pneumatic tires are made of rubber composites reinforced with high strength steel cord. Rubber is cured by the vulcanizing reaction with sulfur addition, and the rubber changes its form from viscous liquid to elastomer, as invented by C. Goodyear in 1839. Rubber polymers are then cross–linked with sulfur in the presence of zinc oxide as a vulcanization activator, forming disulfide bonds between molecular chains. The addition of carbon black to rubber makes the toughened structure, which improves the material properties including wear resistance and degradation resistance for ultraviolet light. This is because rubber polymers are strongly adhered on the surface of carbon black, and the polymer bridges the carbon black particles.

Pneumatic Tire

The pneumatic tires have an advantage of reducing vibrations transmitted from an uneven road surface. However, air pressure may cause the burst of tires in case that the tensile stress of tires exceeds the strength of materials, and the rubber composites are reinforced with steel cord of high tensile strength. The tire pressure is controlled within suitable levels (i) to avoid the burst of tires, (ii) to reduce vibrations for comfortable driving, (iii) to enhance the braking performance, and (iv) to reduce the rolling resistance. Low tire pressure is beneficial for reducing vibrations because the vibrations are transmitted through tires and suspensions, and also for excellent braking performance due to high friction of low–pressure tires. Accordingly, the tire pressures of recent passenger cars are kept around 0.2 MPa, which is considerably lower than that used in the dawn of the motor age. Instead, the cross section areas and the widths of tires are getting large to compensate the low tire pressure.

Friction

The friction of tires strongly depends on weather conditions. The presence of water between the tire and the road surface strongly decreases the friction, and the grooves on the surface of tires are designed to easily eliminate the water from underneath the tire.

The preferred rubber materials are to maintain high friction even under wet conditions. The recent advance is the addition of silica powder to rubber, and the silica addition is found effective for cold weather because the silica containing rubber maintains softness in low temperature resulting in enhancement of the braking force. In addition, the low rolling resistance of the silica containing rubber tire contributes to the reduction in the fuel consumption rate.

Rubber

Natural rubber is an elastic hydrocarbon polymer that is originally derived from latex found in the sap of some plants. The latex is a milky colloidal suspension, and the natural rubber is elastomer, which behaves large elastic elongation.

Large elastic elongation of rubber results from the molecular structure. Rubber polymers are thermally activated even at room temperature, and the movements of cross–linked molecules allow enormous elongations. The elongated molecular structures under tensile loading are highly textured while the molecular structures tend to be random with the removal of the loads. This indicates that the rubber molecules shrink under the free of loads while the tensile stress greatly stretches rubber molecules exhibiting an enormous elongation. Polyisoprene is a purified form of natural rubber.

Steel Filaments

High strength is required for the steel cord reinforcing tires, and the strength is endowed by the fine texture of pearlite that is formed during the extrusion process. Note that pearlite has a lamillar microstructure comprising of ferrite and cementite. The diameter of steels becomes small during the extrusion process, and the pearlite texture also becomes fine as a result of the extrusion process. The fine textured pearlite is responsible for the very high strength of extruded steel filaments.

3.6.2. Brake Discs

A brake is used for reducing the speed of a vehicle and transfers the kinetic energy to frictional heat. The disc brake is advantageous in lightweight and efficient cooling ability in comparison with the drum brake.

Cast Iron Brake Discs

Common cast iron is gray cast iron, which is in a ternary Fe–C–Si alloy in a composite form of dispersed graphite particles in a matrix of ferrite and pearlite, and is characterized with excellent resistance to deformation and wear. Cast iron is produced with the solidification of liquid iron normally containing 2.8 to 3.4mass% carbon and 1.5 to 2.5mass% silicon. Tensile strength of cast iron is governed by the configuration of graphite particles that behave similar to the presence of flaws and cracks, and the brittle behavior of cast iron results from the plate–like shape of graphite particles. Spheroidal graphite particles are produced with the addition of Mg to cast iron, and the spheroidal graphite cast iron exhibits higher strength and elongation in comparison with common gray cast iron. Automotive brake discs are commonly produced from gray cast iron.

Carbon Ceramic Brake Discs

Brake discs benefit from reduced mass through the replacement of traditional heavy gray cast iron with lightweight ceramic materials. Carbon fiber reinforced carbon materials (C/C composites) have been used for airplane brake discs because of their excellent braking performance at elevated temperatures resulting from frictional heat, and intensive studies on the application to automobiles have been conducted.

A carbon–ceramic brake disc is made of carbon fiber reinforced silicon carbide composites. Figure 3.13 shows a carbon–ceramic brake disc installed in an automobile. The advantage using carbon–ceramic brake discs is a significant weight reduction around 65% over conventional brake systems of cast–iron brake discs. It should be noted that the carbon–ceramic brakes also maintain high friction even at elevated temperatures, similar to the characteristics of C/C composite brakes. Carbon–ceramic brakes have already been applied in expensive commercial vehicles.

3.6.3. Brake Pad Materials

The brake pads are mounted on a brake caliper to generate high friction contacting brake discs. The fibrous materials are suitable for high friction, and excellent durability at high temperatures is important due to the generation of frictional heat. Chrysotile asbestos resin composites meet the requirements. However, asbestos fiber is hazardous for human health, and non–asbestos fiber phenolic resin composites are used for recent brake pads. Non–asbestos fibers such as aramid fiber pulps and steel fibers are typically used, and the aramid fibers are polymeric fibers developed for possessing rigid molecular structures.

(a) (b)

Figure 3.13. Carbon–ceramic brake disc made of carbon fiber reinforced silicon carbide composites, (a) a whole view, and (b) installed in an automobile.

Chapter 4

ELECTRONICS IN AUTOMOBILES

In 1860, E. Lenoir produced a two–stroke gas engine, in which a spark plug was equipped for ignition. The high voltage electricity that was required for ignition was supplied with an inductor coil, which is connected to a battery. In 1876, N. Otto invented a four–stroke engine, and a small flame was used for ignition. In 1885, C. Benz employed a spark plug ignition system for automotive gasoline engines.

The phenomenon of electromagnetic induction is that the changes in magnetic field induce an electric current in conductors. This can be applied to electric generators because a coil placed in an alternating magnetic field produces alternative electric current, and the alternating magnetic field may be produced with moving a permanent magnet near the coil.

ELECTRIC SYSTEMS

A pulsing direct electric current is produced in a dynamo, in which coils are mechanically rotated under the static magnetic field, and the direction of electric current in the coil is switched with the combination of commutator and brushes. The direct current produced in dynamos is charged in batteries to be used for automotive equipments.

A starter motor system, developed in 1911 by C. F. Kettering, enabled people to start engines electrically without rotating a crankshaft mechanically by hand. Electric power is supplied from a battery, and the battery was charged with an electric generator. A plentiful supply of electric power enabled numerous electric systems, including electric headlights and windshield wipers.

Electric energy charged in a battery system realized the comfortable life on board, such as audio systems, power window systems and power steering systems. In addition, the development in electronic devices enabled the precise and complicated control of mechanical components including engines, transmission and brakes. As a result, safe driving with a low fuel consumption rate was realized.

ELECTRONIC CONTROL

The electronic control of engines such as fuel injection and ignition timing control has realized the exhaust gas purification system, in which the combustion is monitored with oxygen sensors, and an electronic control unit (ECU) controls the quantity of fuel supplied to each cylinder. The electronic control has contributed to the safety systems of vehicles, including the passive and active safety. Sheet belts and airbags contribute to the passive safety, and the active safety is enhanced with antilock braking system, traction control system and electric stability system.

In recent automobiles, a great amount of information is given to drivers. Automotive cameras detect a view around the car, and drivers are visually informed. In addition, the actual road information is transmitted through a radio beacon equipped on the road in Japan, and the exact position of the vehicle is determined on an automotive navigation system with a help of positional data obtained from a global positioning system (GPS). The development in mutual transmission between the automobile and infrastructure enabled the fare charging system at tollgates. The information technologies enabled a convenient and safe driving, and the technology is seemingly developing toward the realization of automatic driving.

4.1. AUTOMOTIVE DEVICES AND SYSTEMS

Automobiles are powered with heat engines, and the combustion of gasoline engines is ignited with spark plugs. Electric current for the spark plug ignition is supplied from car batteries, which are charged with electric generators installed in vehicles. The battery supplies the electric power to a starter motor and the other automotive devices including lighting and air conditioning.

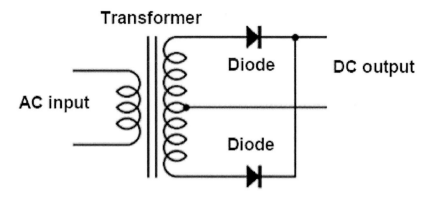

Figure 4.1. Schematic indication of automotive full–wave rectifier using a transformer and two diodes.

4.1.1. Motors and Generators

Automotive Alternators

An electrical generator produces electric power used in automobiles. The electric power is temporarily stored in a rechargeable battery, and is used for the power source of electric devices. A dynamo was originally chosen for the automotive electrical generator. The dynamo produces direct current with the use of a commutator, and the direct current is used for charging batteries.

The development of semiconductor devises has enabled the conversion of alternating current to direct current without using commutator since around 1960. An alternator converts the mechanical energy to the alternating current, and automotive alternators use silicon diodes for rectification. In general, automobile alternators use six diodes comprising a full–wave three–phase bridge rectifier, as shown in Figure 4.1.

ELECTRIC MOTORS

An electric starter is used for rotating the crankshaft to start a reciprocating engine, and the starter motor is connected to the flywheel ring gear. A direct current brushed motor is used for the electric starter, and direct current is supplied from the battery. Numerous small motors are used in automotive devices, including window wipers, power windows, side mirrors and power sheets. The automotive small motors are driven by direct current, and brushed DC motors and brushless DC motors are mostly used. High power motors such as used for power sources of electric vehicles are driven with alternating current, which is produced in a converter unit.

4.1.2. Electric Control of Engines

Ignition Control

The ignition timing was controlled with the mechanical system to detect the rotation of the camshafts, and high voltage was produced in the ignition coil to be supplied to the ignition plugs through distributors. In advanced systems, however, the ignition timing is electronically controlled with a help of angle sensors that indirectly detect the piston position in the engine strokes.

Spark plug ignition initiates the combustion in gasoline engines, and the combustion is completed with the flame propagation. Early ignition timing is thus preferable for high rotational speed to complete combustion within a limited time. The late ignition timing is chosen when severe knocking is apt to occur. The delayed ignition timing is electronically controlled with detecting the vibration due to slight knocking. The slight knocking is detected with piezoelectric ceramic elements in knock sensors to feedback to the electronic control unit.

Fuel Injection Systems

Carburetor was originally used to make air–fuel mixtures, and the fuel was supplied in a form of mist to air. The precise control of the air–fuel ratios became important since then with a technological advance in exhaust gas purification. Fuel injection was employed for precisely controlling the air–fuel mixtures, and the controlled amount of fuel is injected to the intake port. The amount of injected fuel is electronically determined on the basis of various information regarding engine conditions. The engine conditions are monitored with a variety of sensors. Figure 4.2 indicate the outline of an engine control system, in which sensors are used for collecting the signals regarding engine conditions and the signals are transmitted to the electronic control unit to operate actuators. Sensors for engine control system include air flow meters, cam–angle sensors, crank–angle sensors, throttle–position sensors, water–

temperature sensors, oxygen sensors, air–temperature sensors and wheel–speed sensors.

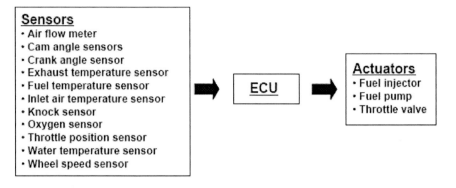

Sensors
- Air flow meter
- Cam angle sensors
- Crank angle sensor
- Exhaust temperature sensor
- Fuel temperature sensor
- Inlet air temperature sensor
- Knock sensor
- Oxygen sensor
- Throttle position sensor
- Water temperature sensor
- Wheel speed sensor

ECU

Actuators
- Fuel injector
- Fuel pump
- Throttle valve

Figure 4.2. Outline of an engine control system, consisting of sensors, electronic control units and actuators.

4.1.3. Lighting

Halogen Lamps

Lighting is an old application of electricity, and the development of electric light bulbs has contributed to high efficient lighting in automobiles. The incandescent light bulb using tungsten filament emits visible light, and light emission is stronger when the temperature of heated substance is higher. The tungsten filament is electrically heated to very high temperature to emit a strong light beam.

The lifetime of the filament is, however, shorter when the temperature is higher. This is because the heated substance is intensively vaporized at elevated temperatures, and the vaporization intensively occurs at the parts of small cross section areas due to the local high electric resistance that produces a great amount of Joule heat. As a result, the vaporization of filament materials determines the lifetime of the bulbs, and the lifetime may be monitored with the darkened bulb surface due to the deposits of the vaporized tungsten metal on the bulb surface.

Longer lifetime was achieved by the invention of halogen lamps, which involve halogen elements such as iodine and bromine in the bulbs. The halogen elements produce the tungsten halides such as WI_6 and WBr_6 on the surface of bulbs, where tungsten metal is deposited. The vapor of tungsten halides is to be deposited on the hot positions to form the tungsten metal. As a result, vaporized tungsten metal, which is deposited on the bulb surface, returns to the original position on the filament, leading to the longer lifetime of halogen lamps. Additional benefit of halogen lamps is to emit strong light because of higher temperature of filaments than that of incandescent light bulbs.

The dimension of halogen lamps is small because the temperature of the bulb surface is required to be maintained around 150°C or higher, enough for promoting the reaction to produce tungsten halides. The temperature of bulb surface is thus much higher than the conventional incandescent lamps, and the bulbs use the high temperature glass such as fused silica and aluminosilicate glass.

High Intensity Discharge Lamps

High intensity discharge lamps emit strong light beams because the temperature attained with discharge is higher than that with resistive heating. The high intensity discharge lamps are similar to mercury lamps, which emit the ultraviolet light resulting from the quantum levels of mercury. The mercury lamps emit the considerable amount of visual light when the pressure of mercury vapor is high enough. Interaction between the activated vapor species reduces the energy of photons, and ultraviolet light is changed to the low energy state of visual light. The high voltage mercury lamps are used in outdoor lighting and emit bluish white light. Note that another discharge lamp is a sodium vapor lamp, which emits yellow light, and used for lighting expressways. High intensity discharge lamps contain xenon gas, the vapor of mercury and metal halides, and white light beam is generated due to the addition of metal halides.

The bulb materials for high intensity discharge lamps require the excellent corrosion resistance against active metal vapor, and fused quartz and translucent alumina are used. Fused quartz is pure silica glass used for mercury lamps, and translucent alumina is made of polycrystalline alumina ceramics, in which porosity is extremely low and light beam is able to penetrate without scattering by pores. The translucent alumina tubes are used for sodium vapor lamps and high intensity discharge lamps due to excellent corrosion resistance to the sodium vapor.

4.1.4. Power Steering

Power steering is a device that assists the power of drivers. In general, small power is assisted with small motors while large power is assisted with the oil pressure mechanism. Actually, the antilock braking system is controlled with the oil pressure, and small motors are used for low powered devises. The power steering system is situated in between, and both systems of oil pressure and electric motor are used.

The rotation of engines is used for rotating the drive shaft resulting in wheel rotation, and additional power is used for rotating a variety of auxiliary machines, such as electric generators, compressors for air conditioning, water pumps, oil pumps and power steering pumps. Oil pumps are used for circulating engine oil, and water pumps are used for circulating cooling water in water jacket to remove heat. The steering fluid, which is pressured with the power steering pump, assists the hand power of the driver to operate steering wheels.

In electric power steering systems, electric motors are used for assisting the power for steering. The electronic hydraulic power steering uses an electric motor for pumping hydraulic fluid, which assists the steering force.

4.2. EXHAUST GAS PURIFICATION

The photochemical reaction of NO_x produces strong oxidants such as ozone and peroxyacetyl nitrate in strong sunshine. Photochemical smog was first reported in Los Angeles in 1940, and it was frequently reported in the early 1970s in the Tokyo metropolitan area. Intensive study was conducted for technology developments to purify exhaust gases from automotive engines because photochemical smog occurs most prominently in urban areas that have large numbers of automobiles. Other requirements are to reduce particulates and farmable gases such as CO and hydrocarbon.

4.2.1. Gasoline Engines

Figure 4.3 shows the composition of the emissions from gasoline engines, indicating that the fuel burns incompletely in the presence of excess fuel, and the resultant gas contains considerable amounts of CO_2, CO, H_2, and hydrocarbons. On the other hand, in the presence of excess air, the exhaust gas is composed of CO_2 and O_2 with small amounts of flammable gases present. The flammable gases remain even after combustion because the flame propagating from ignition is extinguished in the vicinity of the cylinder walls. The content of NO_x remains at a very low level but depends on the reaction temperature. The content of NO_x is therefore highest in stoichiometric combustion, which has the highest reaction temperature.

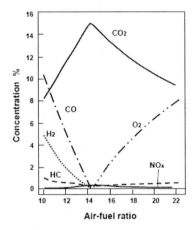

Figure 4.3. Chemical composition of exhaust gas from engines as a function of air/fuel ratio.

Catalyst

Three–way catalyst systems were developed to eliminate NO_x and flammable gases from exhaust. In the absence of oxygen, the flammable gases react with NO_x in the presence of a catalyst, and NO_x is reduced to nitrogen. Therefore, the air–fuel ratio required for purifying exhaust gases is controlled within the limited range corresponding to stoichiometric combustion as shown in Figure 4.4, and oxygen sensors are used to control the air–fuel ratio.

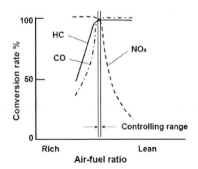

Figure 4.4. Conversion rates of exhaust gas purified with three–way catalyst.

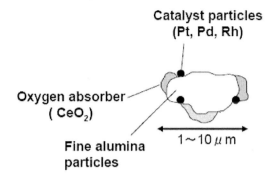

Figure 4.5. Schematic indication of catalyst structure supported on the surface of the cordierite honeycomb.

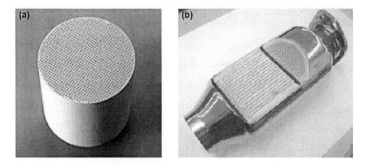

Figure 4.6. Cordierite honeycombs for alumina–supported precious metal catalysts, (a) a photograph of a typical honeycomb, and (b) Cross section of a honeycomb installed in an exhaust pipe of automobiles (courtesy of NGK Insulators, Ltd.).

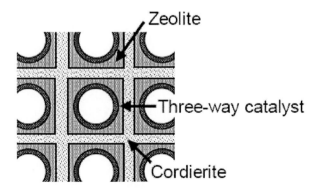

Figure 4.7. Schematic indication of an advanced catalyst system involving a zeolite layer formed on the surfaces of the cordierite honeycomb cells, and the top surface layer is the three–way catalyst.

Chemical reactions occur on the surfaces of the catalyst, which consists of very fine particles of Pt, Rh and Pd as shown in Figure 4.5. The particles are dispersed on the surface of small–grained alumina powder placed on a cordierite honeycomb, as shown in Figure 4.6. The three–way catalyst system starts to operate when the honeycomb catalyst becomes hot. The honeycomb must therefore have thin walls so that it heats up rapidly. Note that the three–way catalyst system does not work immediately after the engine is started because the catalyst is not hot enough to promote the chemical reactions. At this time, hydrocarbons are released in the exhaust gases. An advanced catalyst system involving a zeolite layer for trapping hydrocarbons on the surfaces of honeycomb substrates is shown in Figure 4.7. Hydrocarbons are trapped in the adsorption layer when the catalyst is cold and released when the temperature rises. The released hydrocarbons are purified in the three–way catalyst layer.

4.2.2. Diesel Engines

In diesel engines, oxygen is always in excess, and controlling NO_x emissions by reacting them with hydrocarbons is extremely difficult because the amount of hydrocarbons is much smaller than in gasoline engines. The addition of reducing agents is effective for NO_x decomposition. In the urea–selective catalytic reduction (SCR) system, urea solution is injected into the exhaust pipes, and NH_3 produced as a decomposition product of urea reacts with NO_x to yield N_2 on the catalyst. If NO_x is temporarily adsorbed, the reducing agent provided from temporarily running the engine under rich conditions can react with NO_x. Typically, a $Pt/BaO/Al_2O_3$ catalyst is used to trap NO_x under oxidizing conditions as adsorbed "nitrate" species.

Common Rail Systems

Another problem in diesel engines is particulate generation. Common rail diesel systems were developed to suppress the generation of particulates by reducing the particle size of fuel mist with high–pressure fuel injection. Figure 4.8 shows the common rail system, which comprises of a high–pressure pump for fuel, a common rail for storing high–pressure fuel, and fuel injectors. Fuel is pumped up to a high pressure and temporarily stored in the common rail. The fuel injection is adequately controlled with a control unit to achieve several injections in a single combustion cycle. Particulate generation is reduced by the very fine fuel particles resulting from the high–pressure spray, and reduced NO_x generation results from the gradual combustion due to split injection.

Fuel injection into the combustion chambers is controlled with opening and closing injector valves, and the fuel injected into the chamber starts to burn immediately. For the injection valves, a rapid response is very important because several strictly controlled injections are required in a single combustion cycle. As a result, piezoelectric injectors, which enable a high response, are increasingly being used in place of solenoid valves.

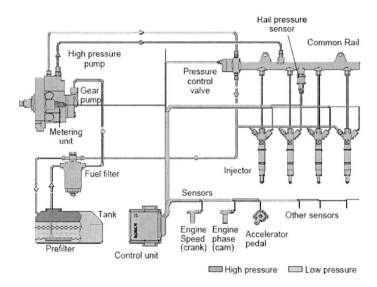

Figure 4.8. Common rail diesel system (courtesy of Robert Bosch GmbH).

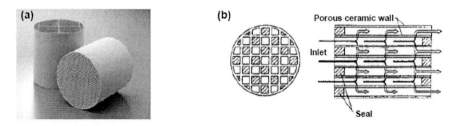

Figure 4.9. Diesel particulate filters, (a) left: silicon carbide DPF and right: cordierite DPF, and (b) schematic indication of the internal structure (courtesy of NGK Insulators, Ltd.).

Diesel Particulate Filter

Even when piezoelectric ceramic injectors are used for common rail systems, particulates are generated during combustion. Some of the particulates are oxidized in an oxidation catalyst while the rest remain even after passing through the catalyst layer. Some diesel particulate filters for finally purifying exhaust gas are shown in Figure 4.9. Exhaust gas is filtered through the porous ceramic walls, and the particulates are captured on the wall surfaces.

4.3. Safety Systems for Vehicles

4.3.1. Passive Safety

Safety systems in automobiles are basically intended to protect passengers and pedestrians from traffic accidents, and systems such as seat belts and airbags may mitigate the severity of the accidents. At the moment of the car crash, the velocity of the automobile is suddenly reduced, and the passengers suffer the force to move forward due to the inertia force. Absorption of the energy of a collision effectively reduces the impact force to passengers, and crumple zones are usually placed in a front of the automobile. The structures composing the crumple zone severely deform during the car crash, and the impact force to the passengers is reduced. On the other hand, the cabin structures are required to be rigid during the crash to protect passengers. The cabin components are built with rigid structures, and high strength steels are used for the components such as pillars and side impact protection beams. The PP composite bumpers are used for mitigating the impact to pedestrians.

Airbag Systems

The strong impact force during the car crash may injure the passengers, and sheet belts supports the passengers to reduce the impact force otherwise the head of a car driver may be strongly damaged due to the collision between the head and windshield. Airbags inflate to cushion the impact of the passengers, and the collision damage of the driver to the steering wheel is, in particular, reduced. In addition, collapsible steering columns reduce the damage to the driver in a frontal crash, and a laminated windshield protects the driver from penetrating the shield because it remains in one piece when impacted.

An airbag system is used for protecting passengers in an automobile collision from striking hard interior objects such as steering wheels. The airbag system consists of sensors to detect collision, an electronic control unit for computing, and an igniter to inflate the airbag. The igniter initiates a rapid chemical reaction generating nitrogen gas to fill the airbag. The accelerometer is of primary importance for detecting the car crash, and the miniature accelerometers with high reliability are used.

4.3.2. Active Safety

The active safety systems assist drivers to reduce the possibility of actual accidents. The good visibility is helpful to drive safely, and easily recognizable turn signals and stop lamps are helpful for oncoming vehicles. The recent advances in the active safety systems are automatically to operate, and a variety of driver assistant systems have been developed.

Antilock braking systems automatically adjust the braking power to effectively reduce the vehicle speed without locking tire wheels. Traction control systems also automatically adjust the rotating wheel to avoid slipping, and the spinning wheels are minimized with a help of electronic stability controlling systems.

Antilock Braking Systems

The antilock braking system is a safety system, which prevents the wheels from locking while braking. Rotational speed of each wheel is monitored with a wheel speed sensor, and the locking of a wheel is detected when the wheel is rotating significantly slower than the others. When the locking is detected, the ECU actuates the valves to reduce hydraulic pressure to the brake at the affected wheel, thus reducing the braking force on that wheel. The wheel speed sensors also contribute to the traction control system, which senses the spinning of the driven wheel during vehicle acceleration and limits power to the drive wheels to minimize the slipping.

Stability Control

The electric stability control (ESC) system is an advanced safety system for stabilizing the driving automobiles by detecting and preventing skids. This basically incorporates yaw rate control into antilock braking and traction control systems. Yaw is rotation around the vertical axis, and the rotation rate of the car is determined with a gyroscope or a yaw rate sensor. The ESC system measures the direction of the skid, and then applies the brakes to individual wheels asymmetrically in order to create torque about the vehicle's vertical axis, opposing the skid and bringing the vehicle back in line with the driver's commanded direction. Additionally, the system may reduce engine power or operate the transmission to slow the vehicle down.

The development of electronic devices is closely related to the advances in sensors and actuators. In particular, reliable sensors are important for the electronic control.

4.4. INFORMATION TECHNOLOGY

Information technology has been recently introduced to automobiles, such as keyless entry system, car navigation and automatic fare charging at tollgates on expressed ways. Key technologies are high–speed computers, wireless transmission including antenna, and sensing devises such as optical cameras and millimeter–wave radars.

4.4.1. Car Navigation Systems

The GPS uses satellites that transmit precise radio wave signals, and the signals received from several satellites determine the exact position of the receiver on the earth. Automotive navigation system uses the GPS information from satellites and is additionally assisted by the beacon installed on the roads. The position of the automobile is thus determined on the map installed in the computer.

The miniaturized GPS antenna is important for realizing the car navigation system in the limited capacity of automobiles, and the patch antenna, which is a variation of micro–strip antenna, is suitable for detecting the electromagnetic wave of microwave frequency. The ceramic patch antenna uses a material of a high dielectric constant between metal plates, and this enables the antenna size smaller. This is because the wavelength is shortened in the substance of high dielectric constant.

A variety of information, such as traffic jam, parking availability and traffic regulation, is now available in the car navigation system in Japan since 1996. This uses vehicle information communication system (VICS), which is based on the infrared beacon on major ordinary roads, radio wave beacon on expressways and FM multiplex broadcasting.

4.4.2. Semi–Automatic Driving

Automatic driving has been partially realized in airplanes and railroads in the case that traffic conditions are stable with very low possibilities of emergency cases. Automobiles are, however, driven in complex situations, and today's automatic driving systems are unfortunately less reliable than manual driving especially in cases of somewhat unusual conditions.

The automatic driving requires the developments in two classes of technologies; one is to sense the circumstances around the vehicles and another is to control vehicles completely. Optical cameras are the typical sensing devises, and image analyses directly contribute to the advance in automatic driving. Information transmitted from the roadside may be helpful for automobiles to judge the circumstances around the vehicles.

Visual monitors assist the view of drivers. A back–view monitor helps the driver when driving backwards, and side view monitors and front view monitors enable to provide the helpful view around the vehicle. Ultrasonic radar systems alert the driver when the vehicle is very close to the obstacles. In advanced systems, images are presented on a computer display as a synthesized image taken from several cameras.

Millimeter–wave radar and infrared laser systems enable semi–automatic driving, in which the constant distance is kept to the foregoing vehicle, and this system has been commercialized as autonomous cruise control system.

Small cameras installed in automobiles are used for detecting white lines painted on the road, and the computer analyses the image to alert the driver when the direction of driving is about to cross the white line. Accordingly, lane–keeping drive is assisted.

Infrared night vision systems are able to detect infrared light emitted by the substance, of which the temperature is higher than the surrounding, and they are used for detecting pedestrians walking on the road.

The vehicle control systems have been developed in safety systems, and a variety of semi–automatic driving systems using optical cameras and radars have been developed. These technologies are in progress and obviously directed toward automatic driving to reduce the possibilities of traffic accidents and traffic jams, although the present state is in assisting and alerting the driver.

In addition, the automatic driving may be advanced in a combination with car navigation systems and transmission technology to communicate with roadside infrastructures.

Chapter 5

FUNCTIONAL MATERIALS AND DEVICES

In the dawn of automobiles, the use of electricity had been very limited, and ignition plugs were the only the parts working with electric power. In the early 20th century, electric power was supplied to the motors for starting engines and to electric bulbs for lighting, and the electric power was produced in electric generators to be charged in lead–acid batteries. The application of electric devices has increased since then and a wide variety of devices have been introduced to automobiles, including a variety of electric motors, sensing devices, and electronic controllers of automobiles.

Electric Devices
Most of the electronic devices used in automobiles are exactly the same as other applications, and numerous numbers of common devices, including resistors, capacitors, inductors, diodes and transistors, are used in the electronic circuits of automobiles. Electronic ceramics and silicon micro–electro–mechanical systems (MEMS) are used for sensors and actuators, most of which are specially designed for automotive applications.

Advance in Electric Devices
Small electric motors are commonly used for automotive actuators, and the electromagnetic valves are used for controlling the flow of fluid such as oil and fuel. The method of controlling actuators depends on the advance in sensing devices and computer units, and the advanced control has been enabled.

Electric motors are switched by hand in the first stage of application. The motors connected to the devices start to work when the switches are turned on by hand, as represented by window wipers, power windows, electric-driven mirrors and power seats.

In the second stage, switches were automatically controlled with computer units that equip sensing devices. In the purification system of exhaust gases, the fuel injection is controlled with the computer unit, and combustion condition is monitored with oxygen sensors.

The third stage of car electronics is in progress. A great amount of data collected on the road conditions are analyzed within a limited time for assisting drivers.

5.1. ELECTRIC MOTORS AND MAGNETIC MATERIALS

Electric motors convert electrical energy to mechanical power with high efficiency, and the use of electric power is associated with the development in the capacity of car batteries.

In the late 19th century, gasoline engines were ignited with flame and heated metal bars, while electric cars use motors for driving. In the early 20th century, car batteries supplied the electric power to automobile starter motors and electric bulbs. Numerous small motors are now available in automobiles represented by electric wiper motors, electric window motors, and side mirror motors, and the use of electric power is increasing with the advance in car batteries and sensing devices.

5.1.1. Electric Motors

The electric motors, consisting of a rotor and a stator, generate mechanical power from electrical energy. A common DC motor is a brushed DC motor that is comprised of a rotating electromagnet in a static magnetic field produced by the stator. The static magnetic field is produced by permanent magnets and electromagnets. Universal motors have similar structure to the brushed DC motor while powered by alternative current.

Another class of motor employs the rotating magnetic field produced by stators, and the magnetic field is mostly followed with coil rotors and permanent magnet rotors while electromagnets and variable reluctance are also available for the rotors. In typical induction motors, a metallic cage rotor follows to the magnetic field by means of electromagnetic induction. In brushless DC motors, the rotating magnetic field is produced with switching the current by means of a controller unit, and the permanent magnet rotor follows the magnetic field. Variation in electric motors is listed in Table 5.1.

Brushed DC Motors

Brushed DC electric motors are a small type of motors and widely used in automobiles because of low production cost. Electromagnets are used for a rotor placed in a static magnetic field, and the direction of current through the coil of the electromagnet is periodically reversed with the combination of a carbon brush and a split ring commutator.

Table 5.1. Electric motors for automobiles

(a) Electromagnet rotor equipped with brush and a commutator

Stators	Motor type	Applications
Permanent magnet	Brushed DC motors	Window wipers, door mirrors and power sheets
DC driven electromagnet	Brushed DC motors	Engine starters
AC driven electromagnet	Universal motors	(Vacuum cleaner, food mixer, and power tools)

Blanket indicates other application than automobiles.

(b) Rotating magnetic field produced by electromagnet stators

Rotors	Motor type	Applications
Electrically conducting coil	Induction motors	(Appliances, air conditioners, and refrigerators)
Permanent magnet	Brushless DC motors	Fan motors for air conditioning (Hard disc drives and CD/DVD players)
	Permanent magnet synchronous motors	High power drive of electric vehicles and hybrid electric vehicles

Blanket indicates other application than automobiles.

However, the mechanical contact between the brush and the commutator is difficult to maintain at high rotating speed because brushes may bounce off the commutator. In addition, the wear of the commutator may require the replacement. The brushed DC motors are suitably applied to the parts occasionally operated, such as starter motors and power window motors. Most of small automotive motors usually employ the permanent magnet for stators, and high performance magnets are used for powerful motors. Electromagnets are also used for the stators, and the starter motors employ the electromagnets stators.

Universal motors are the exactly same type as the brushed DC motors. The brush and commutator system actually work under alternative current, and the universal motors are used for electric vacuum cleaners, food mixers and power tools, in which electromagnets are commonly used for stators.

Brushless DC Motors

Brushless DC electric motors are synchronous electric motors powered by the direct–current power source, and electronically controlled commutation system is installed, instead of a mechanical commutation system based on brushes. In the brushless DC motors, electromagnets in the stator produce a rotating magnetic field, which is followed by the rotor of a permanent magnet. The magnetic field is produced with supplying the electric current of periodically reversing the direction to the electromagnets, and the direction is synchronized with the position of the rotor. The rotor position is monitored with of the Hall–effect sensors, and the direction of electric current for electromagnets is controlled with electronic circuits using diodes.

The brushless DC motors are, in general, expensive because the electronic controller has to be installed. The high efficiency and long lifetime are the major advantages, which are useful for parts requiring excellent durability such as in application to cooling fans and electric power steering.

Induction Motors

Induction motors are a type of alternative current motor where the rotor is powered by means of electromagnetic induction. The electromagnets powered by alternative current produce the rotating magnetic field, which induces secondary currents in the rotors. The induced secondary currents interact with the rotating magnetic field, and a rotational motion of the rotor is created.

The rotors are composed of metallic cage, which generates electric current to produce magnetic field under the rotating magnetic field. In addition, magnetic cores are settled in the cage to enhance the power. The induction motors are widely applied to industrial motors due to their rugged construction, absence of brushes and the ability to control the speed of the motor, and have been used for the power sources of electric propulsion vehicles.

Permanent Magnet Synchronous Motors

Motors used in propellant power sources such as hybrid electric vehicles and electric cars require the very high efficiency, and interior permanent magnet synchronous motors (PM motors) have been developed. This is small and powerful in comparison with induction motors because the strong permanent magnets are contained in the rotors. The early types of the PM motors use the permanent magnets on the surface of rotors, and the improved ones use them in the inner parts of the rotors. Burying the magnets inside the motor provides the basis for a mechanically robust rotor construction capable of high speeds since the magnets are physically contained and protected.

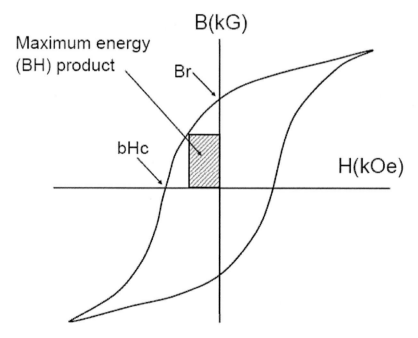

Figure 5.1. *B–H* plot (relation between the magnetic flux density, *B* and the magnetic field strength, *H*) for magnets indicating the residual magnetization, the coercivity, and the maximum energy (*BH*) products.

5.1.2. Magnets and the Related Materials

Most of small motors installed in automobiles use permanent magnets. Power, generated in the motors, depends on the performance of permanent magnets. Ferrite ceramics are extensively used while the advanced rare earth magnets are applied to high performance motors.

Magnetic Properties

Essential requirement for permanent magnets is to maintain the strong magnetic force even under the applied magnetic field. This requirement is realized by the high coercivity and residual magnetization resulting in the maximum energy (BH) product, as shown in Figure 5.1.

Figure 5.2 shows magnetic properties for a variety of magnets commercially available. Note that the magnetic properties vary greatly even within the same class of magnets. It is clear that the residual magnetization is greater in alnico and rare earth magnets than ferrite magnets, and the rare earth magnets exhibits several times greater $(BH)_{max}$ than traditional ferrite magnets. The rare earth magnets are used for high performance motors.

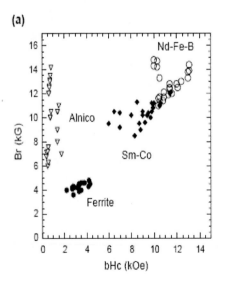
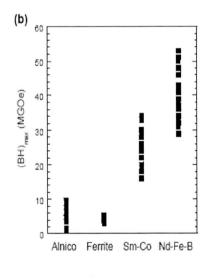

Figure 5.2. Representative magnetic properties for a variety of permanent magnets, (a) residual magnetization and coercivity, and (b) the maximum energy (BH) product. Note that CGS units may be converted from SI units as $1T = 10^4 G$ and $1A/m = 4\pi \times 10^{-3} Os$.

Alnico Magnets

Alnico magnets are primarily composed of iron, aluminum, nickel and cobalt. Before the development of Sm–Co magnets in the 1970s, the alnico magnets were the strongest type of magnets.

Ferrite Magnets

Ferrite magnets are oxide magnets and have high electric resistance among a variety of magnets. The representative ferrite magnets are Ba–ferrite and Sr–ferrite of magnetoplumbite structure, and the chemical formulas are $BaFe_{12}O_{19}$ and $SrFe_{12}O_{19}$, respectively.

The ferrite magnets are the most common among various classes of magnets due to the advantages such as low cost, the excellent resistance to corrosion, the high electric resistance and the high coercivity in comparison with the old types of alnico magnets. The low residual magnetization of ferrite magnets is due to the nature of ferrimagnetism, in which the opposed magnetic moments of the atoms add negative contribution to the net magnetization.

Rare Earth Magnets

Rare–earth magnets are strong permanent magnets made from alloys of rare earth elements, represented by Sm–Co magnets and Nd–Fe–B magnets. Chemical compositions of Sm–Co and Nd–Fe–B magnets are typically Sm_2Co_7 and $Nd_2Fe_{14}B$, respectively. Large residual magnetization results from ferromagnetism. Ferromagnetic materials include the alnico and rare earth magnets, in which all the magnetic moments are aligned in the same direction contributing to the large residual magnetization.

Dysprosium is usually added in Nd–Fe–B magnets to enhance the coercivity, and the powerful magnets are applied to the interior permanent magnet synchronous motors. Regenerative brakes, which use electric motors for regenerating electricity, are used in electric cars and hybrid electric vehicles for efficient power sources.

Electrical Steels

Soft magnetic materials, characterized as small hysteresis areas in B–H curves, are applied to core materials of electromagnets. The additional requirements for the core materials are to have high permeability and low electric conductivity. The high permeability is valuable for producing strong electromagnets, and small hysteresis areas are responsible for high efficiency because hysteresis areas correspond to energy loss. The induction of eddy currents within the core causes a resistive loss, and the high resistance of core materials is useful for reducing the resistive loss.

Pure iron basically meets most of the requirements because of small hysteresis areas and high permeability. High electric resistance of the magnetic cores is achieved both by the modification of materials and the devise structure. The electric resistance of metallic materials is governed by the degree of regular arrangements of constitutive atoms, and the electric resistance of solid solutions is higher than that of pure metals. The silicon electrical steel, which is a steel material doped with silicon, is tailored to meet the requirements for magnetic cores of low electric conductivity. The laminated structure of metals has lower conductivity than solid metals when the current passes across the interface. The electrical steel is manufactured in a form of cold–rolled strips less than 2 mm thick, and the strips are laminated and stacked together to form the magnetic cores.

It is noteworthy that magnetic cores used at higher frequency require lower electric conductivity than that of silicon electric steels, and soft ferrite materials such as MnZn–ferrite and MnCu–ferrite are used. These classes of soft ferrites are crystallized in the spinel structure and exhibit high electric resistance due to oxide materials.

5.2. SENSORS AND ACTUATORS

A variety of sensing devises in recent automobiles are used for automatically controlled systems, such as engines, braking system and airbags. Sensing devises such as wheel speed sensors, accelerometers and yaw rate sensors are used for monitoring the movements in vehicles, and the engine conditions are monitored with various types of sensors including knocking sensors and oxygen sensors. The recent advance in sensor technology may be the development of optical image sensors and radars, which are used for exploring the circumstance around the vehicles.

5.2.1. Si MEMS Sensors

Micro–electro–mechanical system (MEMS) is an integrated micromechanical component produced on a common silicon substrate. The micromechanical components are fabricated through micromachining processes based on the lithography techniques, and parts of the silicon wafer are selectively etched away.

In the process, a silicon single crystal wafer is coated with photosensitive materials, and the mask, in which the fine patterns are described, is placed on the wafer. Photo radiation polymerizes the photosensitive materials, and the fine pattern is transferred to the pattern of polymeric materials formed on the silicon wafer. The etching procedure follows, and the parts of silicon wafer are completely etched away to fabricate the micro–mechanical silicon devises.

Diffusion processes are used for producing p–type or n–type semiconductors, in order to fabricate the electronic circuits on the silicon substrates. The MEMS techniques are applied to the production of accelerometers and the yaw rate sensors.

MEMS Accelerometer

The MEMS accelerometer consists of a thin cantilever beam with installing a gage resistor on the surface close to the fulcrum of the beam. Figure 5.3 shows a representative structure of the MEMS accelerometer or a G–sensor. Once acceleration is applied to the sensor, the cantilever beam bends to create stains at the resistor parts. Note that the gage resistors are made of semiconductor silicon, and the acceleration is monitored with the resistance change in silicon. This is the piezoresistance effect that the electric resistance of the substance depends on elastic strains.

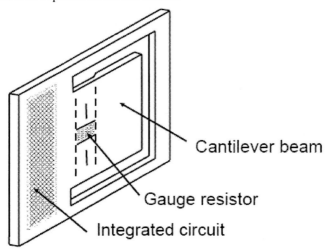

Figure 5.3. MEMS accelerometer.

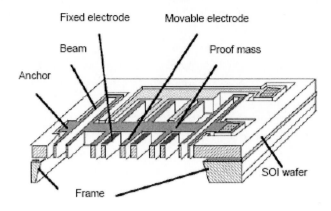

Figure 5.4. Yaw rate gyroscope made of MEMS.

The advanced types of MEMS accelerometers use the capacitance change for measuring the acceleration, and the movement of the loosely fixed part in the accelerometers causes the changes in the gaps between two electrodes for capacitance measurements.

Yaw Rate Sensors

The yaw rate in automobiles is usually determined by a vibration gyroscope, which has a bar vibrating along a certain plane. When the bar is revolted with the rotation of the car, the vibrating plane is also rotated due to the Coriolis force. The yaw rate is determined with detecting the rotation in the vibrating plane.

Figure 5.4 shows a yaw rate sensor of silicon MEMS structure, in which vibration is activated in an electrostatic way and the rotation of the vibrating plane is detected with a change in capacitance. Another candidate for yaw rate sensing is to use piezoelectric materials such as PZT ceramics and single crystalline quartz because piezoelectricity enables to activate and detect vibrations.

5.2.2. Magnetic Application to Sensors and Actuators

Electromagnetic induction is applied to sensors and actuators. Changes in the magnetic field are detected with electric current generation in coil placed in the magnetic field, and the mechanism was applied to wheel speed sensors. Electric current penetrating coil produces a magnetic field to pull a piece of iron, and this movement is applied to the actuator in electromagnetic control valves.

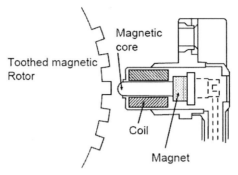

Figure 5.5. Wheel speed sensor of an electromagnetic pick–up mounted on a toothed magnetic rotor.

Wheel Speed Sensors

Wheel speed sensors are used for antilock braking and traction control systems. Figure 5.5 shows a typical wheel speed sensor comprises of an electromagnetic pick–up mounted on an indexing disc connected to the rotating wheel. The index disc consists of a toothed magnetic rotor, and the pick–up contains a permanent magnet and coil for detecting the change in the magnetic field. The voltage generated in the coil varies as the revolution speed of rotors, and the revolution speed is calculated from the frequency of the alternative voltage, which is proportional to the rotation speed. The same mechanism is applied to crank angle sensors and cam angle sensors.

Electromagnetic Control Valves

Solenoid valves are used for switching the flow of fluid, such as fuel injectors and oil pressure control of transmissions, steering and brakes. The valve is controlled with the magnetic force of the solenoid to attract a piece of steel, and the movement of the steel piece switches the flow.

5.2.3. Ceramic Sensors and Actuators

A variety of ceramic sensors and actuators are used in automobiles. Thermistors are composed of metal oxide ceramics and exhibit a large change in electric resistance proportional to a small change in temperature, and are used for sensing temperatures of water, oil and gases. Piezoelectric ceramics are used for knocking sensors and fuel injectors, and zirconia ceramics are used for oxygen sensors.

Thermistors

NTC thermistors are characterized as a negative temperature coefficient of electric resistance and are commonly used in automobiles. The thermistors are fabricated via a sintering process from the powder mixture of transition metal oxides such as manganese, nickel, cobalt and iron. The negative temperature coefficient is due to the semiconducting behavior of the materials. A great number of electrons are thermally activated at elevated temperatures, and the electric conductivity increases at high temperature due to the increased numbers of carriers.

Figure 5.6. Knock sensors of non–resonance type (courtesy of NGK Spark Plug Co., Ltd.).

Knock Sensors

Knock sensors made of piezoelectric ceramic materials are able to detect slight vibrations that occur prior to severe knocking. This enables the engines to operate safely without severe knocking because the ignition timing is delayed with electronic engine controllers to mitigate the combustion. Consequently, knock sensor systems enable gasoline engines to operate in efficient conditions.

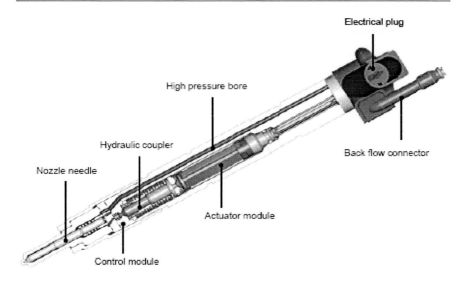

Figure 5.7. Piezoelectric ceramic injector (courtesy of Robert Bosch GmbH).

The knocking sensor consists of a moving weight and a piezoelectric ceramic material, and the piezoelectric element generates high voltage when the weight hits the piezoelectric element. Figure 5.6 shows some non–resonance knock sensors. Slight vibrations caused by knocking are detected with piezoelectric ceramic elements connected to weights that amplify the vibrations in a non–resonant manner.

Piezoelectric Injectors

In a piezoelectric injector, the dimensional changes in the actuator module allow the needle valves to open and close, as shown in Figure 5.7. The main advantages of the piezoelectric injectors are the fast switching speeds and precise fuel metering in comparison with solenoid injectors, and the piezoelectric ceramic actuators are used for fuel injectors of common rail diesel engines and direct injection gasoline engines.

Oxygen Sensors

Solid electrolyte is applied to oxygen sensors for monitoring the oxygen partial pressures of exhaust gases. Oxygen ions diffuse in stabilized zirconia ceramics at elevated temperatures, and the oxygen pressure is monitored with electric voltages resulting from the differences in oxygen partial pressure between both sides of the electrolyte. Figure 5.8 shows the principle of oxygen sensors. The sensor output is governed by the Nernst equation and depends on the oxygen partial pressure of the exhaust gas. Figure 5.9 shows an output voltage of oxygen sensors, and the output is high when the combustion occurs in fuel–rich compositions. Figure 5.10 shows a photograph of oxygen sensors, designed for installing in exhaust pipes.

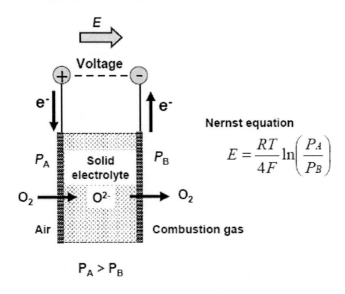

Nernst equation

$$E = \frac{RT}{4F} \ln\left(\frac{P_A}{P_B}\right)$$

Figure 5.8. Principle of oxygen sensors. Zirconia ceramics are used for solid electrolyte, and the output voltage is generated in accordance with the Nernst equation, based on the difference in oxygen partial pressure between air and combustion gas.

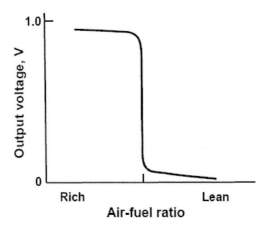

Figure 5.9. Output voltage from oxygen sensors as a function of air/fuel ratio. Output voltage is high in fuel rich combustion conditions.

Figure 5.10. Oxygen sensors made of zirconia ceramics (courtesy of NGK Spark Plug Co., Ltd.).

5.3. BATTERIES AND FUEL CELLS

Batteries and fuel cells produce electrical energy from electrochemical reactions. Batteries are electrochemical conversion devices that store chemical energy as chemical products and produce electric power in a discharge process. Fuel cells enable to produce electricity directly from oxidizing a fuel. The reactions in batteries terminate in principal when one of the electrode materials inserted in the electrolytes is completely consumed. On the contrary, the reactions in fuel cells continue eternally as long as the reactants are supplied to the cells. Rechargeable batteries are able to make reverse reactions that occur during its use by applying electric current.

5.3.1. Rechargeable Batteries

Volta battery may be the most primitive battery, and two different metal plates of zinc and cupper are inserted in an electrolyte consisting of water solutions of acids, bases or salts, as shown in Figure 5.11. In this case, zinc metal is oxidized and solved in the electrolyte solution, and hydrogen gas is generated on the surface of cupper. The electric current is produced as a reaction proceeds until hydrogen bubbles completely cover the cupper cathode surface. Improvements were carried out in the cathode materials to suppress the generation of hydrogen bubbles, and inorganic peroxides were used such as MnO_2, $NiO(OH)$ and PbO_2.

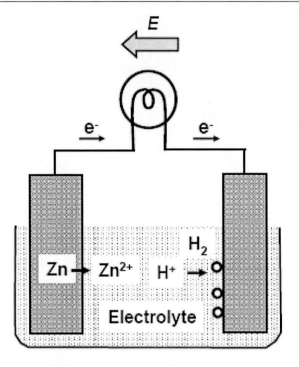

Figure 5.11. Principle of battery, describing a Volta battery.

High Voltage Batteries

Output voltage strongly depends on the selection of anode metal materials while the elements of high ionizing voltage, such as Li, Na, and Ca easily react with water. The elimination of water from the electrolyte is therefore essential for realizing high voltage batteries.

The removal of water from the electrolyte was realized in an organic electrolyte used for the lithium ion batteries and a solid electrolyte for sodium–sulfur batteries. The electrolyte of lithium ion batteries is typically $LiPF_6$ dissolved in carbonate ester, such as ethylene carbonate, and the electrolyte of sodium–sulfur batteries consists of β–alumina ceramics. These batteries are advantageous in high voltage resulting in high power densities

Batteries with Long Lifetime

The lead–acid battery consists of Pb metal anode, PbO_2 cathode and H_2SO_4 for electrolyte. The next generation of the rechargeable batteries was a Ni–Cd battery invented by W. Jungner in 1899, and this uses NiO(OH) for cathode, Cd for anode, and KOH solution for electrolyte. A Ni–metal hydride

(Ni–MH) battery is an improved version of the Ni–Cd battery and use metal hydrides of hydrogen–absorbing alloys, represented by La–Ni, for anode materials instead of Cd.

Even in rechargeable batteries, the number of charge cycles is limited. The life of rechargeable batteries is greatly governed by the morphology change in the electrodes. The electrode materials change their shape during charge and discharge cycles, and the final contacts between two facing electrodes may govern the end of battery life.

Unchanged morphology of electrodes is realized in lithium ion batteries, in which lithium ions and atoms are settled in the atomic sites prepared in the crystal structures during the charge and discharge cycles.

High output voltage and large current are important for electric car batteries, and the internal resistance has to be reduced by placing two facing electrodes closely without contacts. In this view, lithium ion batteries are basically advantageous due to unchanged configurations of electrodes during the charging and discharging cycles.

Lead Acid Batteries

Lead acid batteries, consisting of electrodes of elemental lead and lead dioxide (PbO_2) in an electrolyte of sulfuric acid (H_2SO_4), are the oldest type of rechargeable battery and have been used to provide the current for automobile starter motors.

In discharged state, lead dioxide in the cathode is reduced to lead sulfate ($PbSO_4$), and elemental lead in the anode is oxidized to lead sulfate. As a result, the electrolyte loses its dissolved sulfuric acid. The open–circuit voltage of a single cell is around 2V, and the voltage of car batteries combining six single cells is around 12V.

Ni–MH Batteries

Nickel–metal hydride (Ni–MH) battery is a type of a rechargeable battery and used for hybrid electric vehicles for load leveling. The Ni–MH battery is similar to nickel cadmium batteries but uses hydrogen–absorbing alloy, typically $LaNi_5$, for the negative electrode instead of cadmium. Potassium hydroxide is typically used for the electrolyte.

Hydrogen in the metal hydride in the negative electrode is oxidized to produce water molecule in discharged state, and nickel oxyhydroxide (NiOOH) in the positive electrode is reduced to form nickel hydroxide ($Ni(OH)_2$).

Lithium Ion Batteries

Lithium is a very active element and easily reacts with water. Accordingly, the stabilization of lithium both in electrode and electrolyte was a key for realizing the reliable batteries. The candidates in early stage of developments were TiS_2 for cathode and Li–Al alloys for anode. Li containing non–stoichiometric complex oxides having layered structure, such as $LiCoO_2$, $LiNiO_2$ and $LiMn_2O_4$, appeared as an alterative candidate for cathode materials, and carbon was a new candidate for anode. Li element is able to invade in the vacant lithium sites in the crystal structure of Li containing complex oxides and also to the micro–pores in the carbon molecular structures to form C_6Li.

Negative and positive electrodes in recent lithium ion batteries use the materials of layered structure. Graphite is employed for negative electrodes, and elemental lithium is embedded in the space between graphite layers. Lithium cobalt oxide ($LiCoO_2$) is typically used for positive electrode, while some other candidates of lithium containing complex oxides are intensively considered, including lithium nickel oxide ($LiNiO_2$), lithium manganese oxide ($LiMn_2O_4$), and lithium iron phosphate ($LiFePO_4$). In case of lithium cobalt oxide, lithium ion is inserted into the lithium site prepared in layered structure of non–stoichiometric compound ($Li_{1-x}CoO_2$).

Lithium ion batteries are advantageous for high energy density and applied to the use in electric vehicles and plug–in hybrid electric vehicles.

5.3.2. Electrochemical Double-Layer Capacitors

Alternative candidate of the power source in electric vehicles is an electrochemical double-layer capacitor (EDLC), which is a new type of energy storage device with fast charge/discharge abilities and also called as a super capacitor or an ultra capacitor. EDLC stores the energy in the electric field of the electrochemical double layer at the interface between a solid electrode and liquid electrolyte. Positive and negative ionic charges within the electrolyte are accumulated at the surface of the solid electrode to compensate for the electronic charge at the electrode surfaces. The advantage of EDLC, in comparison with rechargeable batteries, is high power density while energy density is considerably smaller than the rechargeable batteries. Ammonium tetrafluoroborate salts that are dissolved in propylene carbonate and sulfuric acid are used for the electrolyte, and activated carbon is employed for the electrode.

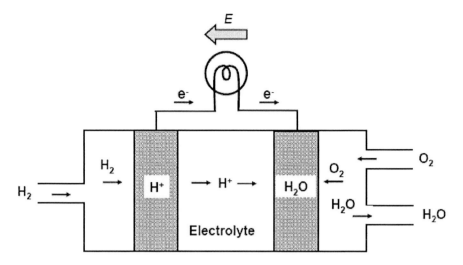

Figure 5.12. Principle of fuel cells.

5.3.3. Fuel Cells

Fuel cells convert chemical energy to electricity similar to batteries as long as fuel is supplied. Fuel cells basically consist of electrodes and electrolyte, and variations come from the selection of electrolyte, as shown in Figure 5.12. A wide variety of electrolyte has been investigated, and the actively investigated electrolyte includes liquid phosphoric acid, proton–conducting polymer membranes, and stabilized zirconia ceramics.

Phosphoric Acid Fuel Cells

Phosphoric acid fuel cells (PAFC) are a type of fuel cell that uses liquid phosphoric acid as an electrolyte. The electrodes are made of carbon paper coated with a fine–dispersed platinum catalyst.

PAFCs have been used for stationary applications with a combined heat and power efficiency of about 80%, and they continue to dominate the on–site stationary fuel cell market. The phosphoric acid fuel cells, commercially available as a cogeneration system generating on–site electricity and hot water, use a liquid phosphoric acid as an electrolyte. Hydrogen is supplied to the surface of anode and is dissolved in the electrolyte as hydrogen ions. The hydrogen ions react with oxygen at the cathode surface to produce water molecules. The electrodes are typically made of carbon paper coated with a fine–dispersed platinum catalyst in order to enhance the reaction around

200°C. Hydrogen is usually formed using the steam reforming reaction of natural gas.

Polymer Electrolyte Fuel Cells

Polymer electrolyte fuel cells (PEFC) use proton–conducting polymer membranes for electrolytes, and are used for fuel cell vehicles. Hydrogen gas is supplied on the anode side, and hydrogen diffuses to the anode catalyst and dissociates into protons and electrons. The protons conduct through the membrane to the cathode, but the electrons are forced to travel in an external circuit with supplying power because the membrane is electronically insulating. The protons arrived on the cathode catalyst form water with reacting to oxygen molecules supplied on the cathode and the electrons, which have traveled through the external circuit. Extensive use of platinum catalysts in electrodes results in very high cost of this class of fuel cells. Another problem is the impurities, which are involved in the hydrogen gas to cause degradation of the catalysts. Metallic impurities also degrade the proton–conducting polymer membranes because metallic ions chemically react to degrade the proton–conducting polymer membranes.

Solid Oxide Fuel Cells

The ceramic electrolyte is used for the solid oxide fuel cells (SOFC). The advantages are high efficiencies, long–term stability, fuel flexibility, and low cost. Y_2O_3–doped stabilized zirconia (YSZ) is typically used for the electrolyte because oxygen ions are able to penetrate in the solid electrolyte. The numerous vacant defect sites of oxygen are present in the materials, and are responsible for the oxygen conductors at elevated temperatures.

It is noteworthy that high operating temperature around 700 to 1000°C enables the use of non–precious metal electrodes. Conductive ceramics are candidates for the oxygen electrode such as Sr–doped $LaMnO_3$, and Ni–YSZ composites are used for the fuel electrode. The greatest disadvantage for automotive application is, however, the high operating temperature requiring long start up times.

5.4. AUTOMOTIVE ELECTRONIC DEVICES

Automotive electronics in general consist of sensing devises, electronic circuits, actuators and batteries. The electronic circuits, consisting of semiconductor devises and other passive electronic parts, are installed on the wiring circuit boards to be placed in the automobiles. The requirements for the electronic devices are thus tough and reliable because the electronic devises may be exposed in severe environments.

5.4.1. Electronic Ceramics

The range of electronic ceramics covers from various functional materials to insulator materials. Former includes a variety of function, represented by piezoelectric ceramics, magnetic ceramics and solid state electrolytes, as described. Automotive application of ceramic insulators includes wiring circuit boards, spark plug insulators, and substrate for high power inverters.

Ceramic Multilayer

Figure 5.13 shows an electronic control unit, where the wiring circuit boards are made with ceramics. This is because the devises are installed in engine rooms, and materials are required to have excellent resistance to high temperatures. Figure 5.14 shows the internal structure of ceramic wiring circuit boards, which consists of ceramic multilayer involving metallic conductors to make electrical contact between the installed electronic parts, such as semiconductors, capacitors and resistors.

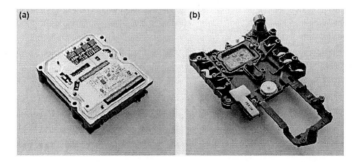

Figure 5.13. Electronic control units using multilayer ceramic wiring board for (a) antilock braking system and (b) transmission control unit (courtesy of Murata Manufacturing Co., Ltd.).

The multilayer are produced from ceramic green sheets, on which metallic wiring patterns are printed, and the ceramic green sheets are produced from the mixture of ceramic powders and organic solvent containing organic binder.

Ceramic multilayer technology is also applied to ceramic capacitors and inductors. The ceramic multilayer capacitors consist of thin dielectric ceramic layer with metallic electrodes on both sides, and metallic coil in a matrix of magnetic ferrite ceramic is formed in the ceramic multilayer inductors.

Spark Plug Insulators

Ceramic materials are also used for insulators of spark plugs, and low soda alumina ceramics are used. The requirement of spark plug insulators is excellent electric resistance at elevated temperatures because the insulator temperature is raised by the spark plug ignition in the vicinity of combustion chambers. The elimination of soda from alumina powder has contributed to the development in spark plugs since the conduction of sodium ion penetrating the ceramics degrades the insulation performance.

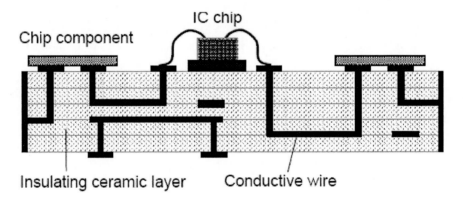

Figure 5.14. Internal structure of multilayer ceramic wiring board.

Substrates for High Power Inverters

The prompt elimination of heat generated in inverters is increasingly important because huge electric current produces the great amount of Joule heat. The structure and materials for promptly eliminating heat are intensively investigated. Basically, materials of high thermal conductivity are suitable for substrate materials.

High thermal conductivity in metallic materials is associated with high electric conductivity, and the suitable choice is a metal of high electric conductivity, such as aluminum and cupper. This is because thermal conductivity is strongly correlated with electric conductivity in metallic materials, and most of thermal energy is transferred by electron. Therefore, Al and Cu are used for heat sink of high power inverters.

Thermal conductivities of insulators are, however, governed by the vibration of constitutive atoms, and high thermal conductivity is achieved in crystals consisting of lightweight atoms having strong chemical bonding, such as diamond, cubic BN, BeO, AlN and SiC. In addition, eliminating impurities and imperfections is important for achieving the high thermal conductivity. AlN ceramics of tailored microstructure are chosen for the substrate of IGBT (Insulated gate bipolar transistor) inverter modules, used for hybrid electric vehicles.

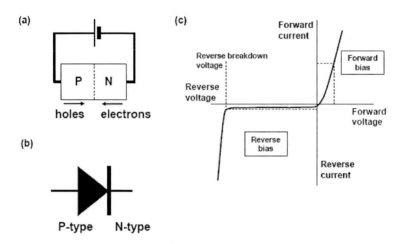

Figure 5.15. Diode, two types of semiconductors are joined together, (a) diode structure indicating forward bias, (b) schematic symbol of diode, (c) *I–V* characteristics of diode.

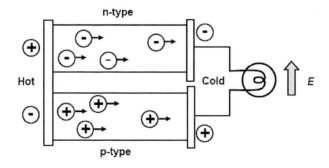

Figure 5.16. Principal of thermoelectricity.

5.4.2. Semiconductor Devices

Silicon semiconductor is used for MEMS sensors while the electronic devices use numerous semiconductor devises such as electronic control units, rectifiers and inverters. In addition, thermoelectric devices are comprised of semiconductor materials.

Diode
Diode is a simple joint of p–type and n–type semiconductors, and electric conductance depends on the direction of applied electric field. Figure 5.15 shows the structure of diodes with a current–voltage characteristic. Electric conduction is high in the direction of forward bias while the opposite direction of reverse bias has very low electric conductance when the voltage is lower than the reverse breakdown voltage. The diodes are used for the devises of rectifiers and inverters.

Thermoelectric Devices
A thermoelectric device directly converts a temperature difference to electrical energy, and the temperature difference is produced with applying electric field to the devise, as shown in Figure 5.16.

Seebeck coefficient is a measure of the magnitude of the induced electric field. Large electric power is generated in a material having high Seebeck coefficient and high electrical conductivity because electric power generated in thermoelectric devices depends on the voltage and current. In addition, a material of low thermal conductivity is advantageous for yielding large thermoelectric power because low thermal conductivity generates a large temperature difference between two ends of the material.

Accordingly, the performance of thermoelectric devices is evaluated with the figure of merit Z defined as

$$Z = \frac{\sigma S^2}{\kappa} \tag{5.1}$$

where σ is the electrical conductivity, κ is the thermal conductivity, and S is the Seebeck coefficient.

More commonly used is the dimensionless figure of merit ZT, in which Z is multiplied with temperature T, and the dimensionless figure of merit is directly related to the thermodynamic efficiency.

It should be noted that thermoelectric materials of high performance have in general high ZT values greater than unity.

Chapter 6

INDUSTRIAL MATERIALS
FOR CREATING THE FUTURE

Well known is the fact that the quality of human life is associated with the development in materials. In the Old Stone Age, ancient humans used cutting tools that are the products of stone in the shape of pebble and flake. In the New Stone Age, agriculture was adopted by humans. Polished stone tools were developed for cutting trees to produce the field of plants, and pottery was fabricated by firing the clay products formed in desired shapes. The technological development of producing high temperatures led to the production of metallic materials, and the Bronze Age began. Bronze is a hard alloy of copper containing tin and related elements, and the melting temperature is 1000°C or lower, depending on the chemical compositions. High temperature is required for producing iron metal in comparison with the bronze production, and the development of high temperature kilns resulted in the beginning of the Iron Age.

The Advance in Industrial Materials

Industrial revolution enhanced the development of steel making. The blast furnace was developed for producing pig iron. The blowers, which were powered by Watt's steam engines, contributed to the construction of blast furnaces working at high temperatures, which enabled large-scale cast iron production.

Nowadays, a great amount of steel is produced, and the converters are used for eliminating excess carbon from pig iron. Oxygen gas is introduced to the molten iron in the container of the converters, and reacts with carbon involved in the melt to produce carbon monoxide gas, which is soon eliminated from the system. Steel sheets are produced from the molten steel. The slab is produced by casting the molten steel, and is hot and cold rolled to the shapes of steel sheets. The steel sheets are then cut and deformed into the desired shapes, and some of them are welded together to produce a complex shape.

Concrete and mortar are the most important application of cement, and the modern cement industry was started in 1824 when Joseph Aspdin invented portland cement. The reinforced concrete structure with steel frame and skeleton enabled the construction of facilities such as tall buildings and bridges. In the 20th century, new classes of materials contributed to the industry, including synthetic polymers, fiber-reinforced plastics, silicon semiconductor devices and advanced ceramics.

Materials in Industries

Clear definition of industrial materials may be difficult because the natural resource changes its shape in each manufacturing step. The natural resources are purified to raw materials having chemical homogeneity, and the purity is important for evaluating the raw materials. The next step is to enhance the properties of materials with modifying the dimensional shape and microstructures. The materials are evaluated with the properties and dimensional preciseness. Devices are produced from combination of materials, and the performance of the devises is the important characteristic. Finally, the devices are combined together for fabricating the final products.

How to Develop New Materials

Numerous advanced materials have been developed and have contributed to modern human life. However, the way to develop adequate materials is not straightforward, and some of them have been occasionally developed in an unexpected way. The way of developing new materials is not always the application of established scientific knowledge, and the new materials are in many cases associated with the discovery of scientific knowledge. Accumulated knowledge is helpful for further improvements in materials such as advanced steels because a great amount of investigation has been already conducted in the field. However, the accumulated knowledge is in most cases insufficient to develop new materials.

The discovery of new material structures includes fullerenes and nanotubes. Buckminsterfullerene C_{60} was discovered in 1985. At the time, structures such as fullerenes are unpredictable, and graphite and diamond structures were known as the crystallized forms of carbon. The discovery of new functions includes high–temperature superconductors and titanium dioxide photo–catalysis. The high–temperature superconductor is a new class of superconductor of cuprate–perovskite ceramic materials discovered in 1986 with a surprise, and many of the superconducting transition temperatures of the high–temperature superconductors are above 77K that is the boiling point of nitrogen. The surprise is due to the consensus at the time among the academic researchers in the field, and the upper limit of the superconducting transition temperatures has been believed around 30K, and ceramic materials have not been considered as a candidate for high–temperature superconductors before the discovery. The photo–catalytic property of titanium dioxide was suggested in the Honda–Fujishima effect, which is an unpredictable phenomenon that titanium oxide electrode produces oxygen gas in an aqueous solution when it is exposed to the strong light. This is a photo–chemical decomposition of water molecule.

Accordingly, the development of new materials is in many cases accompanied with the acquisition of new scientific knowledge, and the material technology has been developed along with the efforts toward two directions of scientific knowledge accumulation and engineering application of the scientific knowledge. It should be noted that the application of scientific knowledge is important and a steady way to proceed the technology. However, the discovery of innovative materials has been associated with scientific advances, and new fields of materials research were started. The appearance of new materials that will greatly contributed to human life in the future is therefore unpredictable and may appear with a surprise.

APPENDIX A: ESSENTIAL CONCEPTS IN MATERIALS SCIENCE

Industrial materials are used for fabricating a variety of devices and final products, and a great amount of steel and cement has been produced for manufacturing mechanical devices and constructing structures. In the 20th century, new classes of materials, such as synthesized polymers, aluminum alloys, advanced ceramics and semiconductors, are added to the industrial materials. Synthesized polymers contributed to the production of fabric and containers. Aluminum alloys realized lightweight airplane bodies. The process of fabricating sheet glasses was greatly advanced, and the production of large glass windows contributed to the daylight design of buildings. Advanced ceramics contributed to electric devices such as alumina insulators, high frequency ferrite core and ceramic capacitors. Semiconductors greatly contributed to electronic devices, and the invention of diodes and transistors has completely replaced the vacuum tubes. The development of integrated circuits realized the computer technology, and further advances are being expected for the technical innovations in the 21st century.

Although numerous valuable references are available on materials science and engineering, ranging a wide variety of materials, processes and properties, the simplified and essential concepts in materials science are summarized in this appendix.

A1. CHEMICAL BONDING IN SOLIDS

Chemical bonding in solids governs the structure and properties of materials, and this is associated with the electron configuration of the atoms.

The electron configuration may be calculated from the Schrödinger equation in the assumption that a single electron is present. A wave function for the single electron is solved to determine the energy state of electron while the problem is in the cases of multi–electrons because the calculation of potential energy between two electrons requires the precise information regarding both position and momentum. The preciseness in the both values is limited by the uncertainty principle proposed by Heisenberg in 1927, and the determination of potential energy between electrons is impossible without some approximation.

Electron Configuration of Atoms

The electron configuration for each element has been calculated, on the basis of the energy state of hydrogen atom. The energy state of electron configuration is usually in the ground state, in which electrons are present in the lowest energy state. As a result, the energy state of electron in atoms is expressed in four quantum numbers; n: principal quantum number, l: azimuthal quantum number, m_l: magnetic quantum number and m_s: spin projection quantum number, as shown in Table A1. Note the quantum numbers of n, l and m_l are variable while m_s has two fixed values of -1/2 and +1/2. Therefore, two electrons are acceptable in each orbital determined by the combination of quantum numbers of n, l and m_l.

Covalent Bond and Metallic Bond

Once the atoms make chemical bond between them, the electron configuration may be changed from the orbital of an isolated atom. The total energy of electrons is reduced due to the stabilized structure in electronic configurations.

Covalent bonding and metallic bonding are strongly associated with electron configuration. In covalent bonding, some electrons are sheared by neighboring atoms in the molecular orbital bridging elements, and the energy level of the orbital are stabilized in the covalent bonding, in comparison with isolated atoms. In metallic bonding, numerous free electrons are sheared in the materials, and the numerous energy levels of free electrons are produced. The energy levels of free electrons are very similar each other and produce the energy band.

Table A1. Quantum numbers and atomic orbital

n: principal quantum number	l: azimuthal quantum number	m_l: magnetic quantum number	Orbital	The number of electrons in the orbital
1 (K–shell)	0	0	1s	2
2 (L–shell)	0	0	2s	2
	1	0, ±1	2p ($2p_x$, $2p_y$, $2p_z$)	6
3 (M–shell)	0	0	3s	2
	1	0, ±1	3p ($3p_x$, $3p_y$, $3p_z$)	6
	2	0, ±1, ±2	3d (five in total)	10
4 (N–shell)	0	0	4s	2
	1	0, ±1	4p ($4p_x$, $4p_y$, $4p_z$)	6
	2	0, ±1, ±2	4d (five in total)	10
	3	0, ±1, ±2, ±3	4f (seven in total)	14

Ionic Bond

Electrostatic energy is another contribution to chemical bonding. In ionic solids, electron transfer occurs between two or more different elements, and positive and negative charged ions are arranged to generate the attraction force. The electrostatic energy between ions stabilizes the crystal structure in ionic crystals.

Van der Waals Force and Hydrogen Bond

Van der Waals force comes from the fluctuations in the electron cloud around an atomic nucleus, and the weak attraction force results from the polarization of electron clouds. The weak attraction force combines the molecules and atoms in solids of organic polymers and rare gases.

A hydrogen bond is the attractive interaction of a hydrogen atom with an electronegative atom. The hydrogen bond is stronger than a van der Waals interaction, but weaker than covalent or ionic bonds. Intermolecular hydrogen bonding is responsible for heat resistant organic polymeric materials and the high boiling point of water and the related compounds.

Figure A1 schematically shows the chemical bonding in solids.

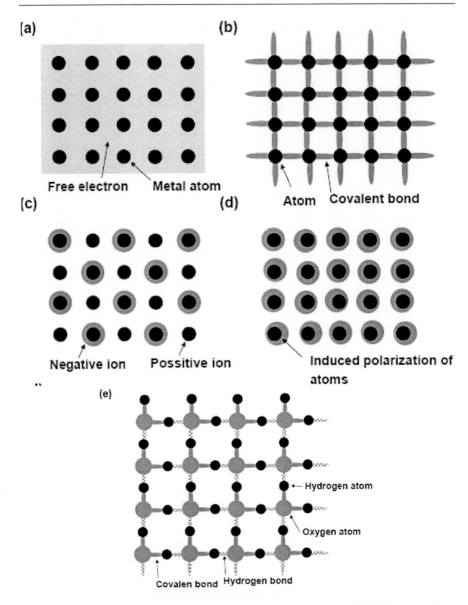

Figure A1. Chemical bonding in solids, (a) metallic bonding, (b) covalent bonding, (c) ionic bonding, (d) van der Waals bond, (e) hydrogen bonding in water molecules. Free electrons are present in the metallic bonds, localized electrons contribute to the covalent bonds, electrons are transferred from positive to negative ions in ionic bonds, the induced polarization of electron cloud are responsible for van der Waals force, and hydrogen bonds the water molecules.

A2. STRUCTURES OF SOLIDS

Metals

In metallic bonded materials, metallic ions of positive charges are regularly arranged in a wide spread electron cloud. The crystal structures of metals are rather simple, and similar to the models of closed packed spheres. Accordingly, most of metallic materials are classified into three types of crystal structures, as shown in Table A2 and Figure A2.

Table A2. Crystal structures of metals

Crystal structures	Metals
FCC	Al, Ca, Ni, Cu, Sr, Rh, Pd, Ag, Ir, Pr, Au, Pb, Th
HCP	Be, Mg, Ti, Co, Zn, Y, Zr, Cd, La, Hf, Re, Os
BCC	Li, Na, K, V, Cr, Fe, Rb, Nb, Mo, Cs, Ba, Ta, W

(a)   **(b)**   **(c)**

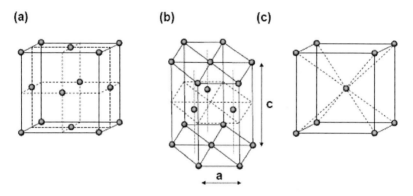

Figure A2. Representative crystalline structures of metallic materials, (a) FCC, (b) HCP and (c) BCC.

Face centered cubic (FCC) lattice is a representative crystal structure in metals, which realizes the crystalline structure of closed packed spheres of metal elements, and the unit cell of FCC lattice is shown in Figure A2(a). In the actual materials, the crystals are formed with numerous combinations of the unit cells, and the material is composed of numerous crystals and secondary phases, involving numerous imperfections such as lattice defects and dislocation in the crystals.

Table A3. Variation in the c/a ratio of HCP metals

HCP Metals	Ratio c/a
Be	1.568
Os	1.579
Hf	1.587
Ti	1.587
Y	1.588
Zr	1.592
La	1.613
Co	1.623
Mg	1.623
Ideal	1.633
Zn	1.856
Cd	1.886

Figure A2(b) shows another closed packed structure of hexagonal closed pack (HCP) lattice, which is similar to the face centered cubic lattice while the stacking order is slightly different. HCP lattice theoretically requires the c/a value of axial ratio to be 1.633 while the actual ratios are slightly different as listed in Table A3, indicating somewhat loosely packed structure. Body centered cubic (BCC) lattice is a dense packed structure shown in Figure A2(c) but has lower packing density than that of FCC lattice.

(a)   (b)

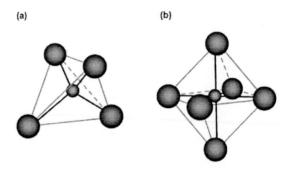

Figure A3. Coordination in ionic crystals, (a) a small positive ion are surrounded by four negative ions in the tetrahedral coordination, (b) large positive ions are surrounded by six negative ions in the octahedral coordination.

Ionic Crystals

In ionic solids, electrostatic energy governs the stability of solids. Positively charged ions are positioned closely surrounded by negative charged ions, and the structure of lowering electrostatic energy is preferable. In this view, the positive and negative ions are closely positioned. However, repulsive force is large when the distance is too close. Accordingly, the suitable distance between two ions is determined. This may be divided to two parts: ionic radius for positive and negative ions. The representative ionic radii are listed in Table A4.

Table A4. Ionic radii of selected elements*

Ion	Radius (pm)
O^{2-}	140
S^{2-}	184
F^-	136
Cl^-	181
Li^+	60
Na^+	95
Cu^+	96
Ag^+	126
K^+	133
Mg^{2+}	65
Zn^{2+}	74
Ca^{2+}	99
Sr^{2+}	113
Ba^{2+}	135
Al^{3+}	50
Ga^{3+}	62
Y^{3+}	93
La^{3+}	115
Si^{4+}	41
Ti^{4+}	68
Zr^{4+}	80

* Pauling, L., *The Nature of the Chemical Bond, 3rd ed.*, Ithaca, New York, Cornell Univ. Press, 1960.

The ionic radius of a positive charged ion is smaller than that of negative charged one because electrons are added to the negative charged ions taken from the positive charged ions. Accordingly, the number of negative ions surrounding the small positive ion is governed by geometrical configuration, and the ionic radius ratio determined the number of surrounding negative ions. Figure A3 shows the examples of atomic coordination. In general, small–sized positive ions such as Si^{4+} are surrounded by four oxygen ions in tetrahedral coordination, and large positive ions such as Al^{3+} and Mg^{2+} are surrounded by six oxygen ions in octahedral coordination.

The range of stable coordination is derived from a geometrical view, and the coordination number is governed by the ionic radius ratio between ionic radius of positive ion R_M and negative ion R_x in the case of simple crystals composed of positive ion M and negative ion X. Accordingly, tetrahedral coordination is stable when the ratio of R_M/R_x is in a range of $0.224 < R_M/R_x < 0.414$, and octahedral coordination is stable when $0.414 < R_M/R_x < 0.732$.

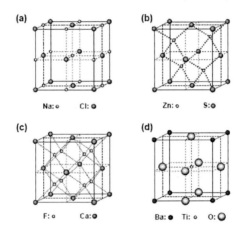

Figure A4. Crystal structures of ionic solids, (a) the rock salt structure of NaCl, in which the Cl^- ions are arranged in the CCP structure and all the octahedral sites are filled with Na^+ ions, (b) the zinc blende structure of ZnS, in which the S^{2-} ions are arranged in the CCP structure and half of the tetragonal sites are filled with Zn^{2+} ions, (c) the fluorite structure of CaF_2, in which the Ca^{2+} ions are arranged in the CCP structure and all the tetragonal sites are filled with F^- ions, (d) the perovskite structure of $BaTiO_2$, in which the Ca^{2+} and O^{2-} ions are arranged in the CCP structure and a quarter of octahedral sites are filled with Ti^{4+} ions.

It is noteworthy that the actual chemical bond in ionic bonding is normally in the mixed covalent and ionic bonding, and the crystal structure is influenced by the covalent contribution to ionic bonding.

Various types of crystal structures are present in ionic crystals. Some of simple crystal structures may be related to the CCP structure of large ions, in which octahedral and tetrahedral sites are filled with small ions. Figure A4 shows the representative crystal structures of NaCl, ZnS, CaF_2, and $BaTiO_3$.

Covalent Materials

Covalent bonds are associated with the formation of hybridized orbital, such as sp^3 hybridisation. Carbon element in organic compounds forms sp^3 hybrid in many cases, and carbon is tetrahedrally coordinated. This is because ground–state configuration of carbon is $1s^2\, 2s^2\, 2p_x^1\, 2p_y^1$, and four sp^3 hybrids are produced from mixing of 2s orbital and three 2p orbitals. Typical example of sp^3 hybrids is diamond, which is composed of carbon, and the carbon atoms are tetrahedrally coordinated each other. The actual crystal structure is the same as ZnS, where all the Zn and S ions are replaced with C.

Silicate is another example forming covalent bonding. Silicate contains a variety of elements, such as Na, K, Mg, Fe, Ca and Al, while the main structure is constructed with silicon and oxygen atoms. In typical silicates, Si is tetrahedrally coordinated with oxygen, and oxygen bridges two silicon atoms.

Note that electrons that contribute to the covalent bonding are present in the orbital between atoms while electrons that contribute to the ionic bonding have been already moved from positive ions to negative ions. Accordingly, the chemical bonding in actual insulating materials may be in the intermediate stage between covalent and ionic bonds.

Molecular Crystals

Chemical bonding due to van der Waals force is very weak while major bonding force in rare gas crystals and molecular crystals. In molecular crystals, covalent bonding is formed inside of the molecules, and van der Waals force works for weak bonding between molecules. Even in molecular crystals, hydrogen bonding between molecules reinforces the structure resulting in enhancements in the thermal and mechanical properties.

A3. MATERIALS

The major structural materials may be classified into metals, polymers and ceramics. Metals are formed with metallic bonding, and high electric conductivity and thermal conductivity are derived from the presence of free electrons. Polymeric materials are covalent bonded in intermolecular structures while the weak bonding between the molecules determines the properties such as low melting temperature, low modulus and high thermal expansion coefficient. Ceramics commonly consist of two or more elements, which are combined each other with ionic and covalent bonds. Ceramics are thus characterized with a low thermal expansion coefficient, high melting temperature and high elastic modulus. Ionic bonding is responsible for hard and brittle behavior in comparison with metallic materials due to the multi–elemental composition, and the unit of the repeated atomic structure in ionic crystals is larger than that of metals.

The materials are designed to have high performance, and the texture and chemical composition are adequately controlled. In addition to the grain size and crystal phases, the distribution of secondary phases and precipitates are typically controlled.

A3.1. Metals and Steels

Metallic materials may be characterized by high electrical conductivity because atoms in a metal readily lose electrons to form positive ions, which are surrounded by the electron cloud of delocalized electrons. The metallic bonding is thus non–directionally formed in nature, and this makes rather simple crystal structures, such as BCC, HCP and BCC. The ductility of metal also comes from the non–directional nature in chemical bonding, and atomic planes in a metal are able to slide across one another under shear stresses.

Steels are the most important structural materials, and the most of vehicles except airplanes are made of steels. Carbon steels are iron alloys containing a small amount of carbon, and the strengths of steels are greatly controlled by the content and distribution of carbon in the materials. In general, carbon precipitates as cementite phase in steels, and the dimension and morphology are controlled with heat treatments.

A variety of steel materials have been developed. Steels exhibiting excellent formability have been developed with reducing the carbon content, and hard and toughened steels are developed with the fine dispersions of hard particles. Cementite particles are typically dispersed in steels while the other intermetallic compounds are dispersed in high alloy steels.

A3.2. Ceramics and Glass

Most of ceramics and glasses are transacted in a form of final products because the materials are difficult to be deformed, machined and joined. The final products of glasses are directly produced from glass melts, and ceramics are produced via a sintering process with minimizing the final machining processes.

The fabrication process of industrial ceramics is as follows. First, fine powder is compacted into the desired shapes with a help of organic binders. The powder compacts are fired for sintering at elevated temperatures below the melting temperature, after the removal of organic binders and solvents. Sintering is a process of densification of powder compacts, and the excess surface energy of each powder particle enhances the densification by means of thermally activated diffusion of constitutional ions and elements.

Glass is a substance of amorphous state, represented by silicate glass. Since the structures of silicates are based on the covalent bonding of SiO_4 tetrahedral units, the regularly aligned structures may be disturbed by the incorporating alkali and alkali earth elements. Therefore, Na_2O and CaO are incorporating in the common soda–lime silicate glass, and no long–range regularity is present in atomic arrangement similar to the liquid structure.

The glasses are produced from the gradual cooling process of melt, and high cooling rate may cause the residual stress responsible for unexpected failure. However, the controlled cooling of glass causes the generation of compression stress on the glass surface, and the stress contributes to the increase in the tensile strength. The tempered glass sheets are produced with quenching at the temperature of 700°C by blowing air. The general concept of strengthening glass is to produce compression stress on the surface.

A3.3. Polymer and Composites

Polymer is a large molecule composed of repeating structural units typically connected by strong covalent chemical bonds while the chemical bond between molecules is very weak, resulting in poor mechanical and thermal properties. Some of polymers are organic compounds produced biologically and others are chemically synthesized. Polymers represented by cellulose, proteins and natural rubber are biologically produced, and most of plastics and elastomer are synthetic polymeric materials.

The synthetic polymer may be classified to thermoset plastics and thermoplastics. Thermoset plastic materials are usually liquid prior to curing, and irreversibly cured into a plastic or rubber by a cross–linking process, represented by bakelite, urea–formaldehyde foam, melamine resin, and epoxy resin. The cross–linking process is promoted by irradiation, heat and a chemical reaction.

Most commonly used plastics are commodity plastics of thermoplastic resin, represented by polyethylene (PE), polypropylene (PP), polystyrene (PS) and polyvinyl chloride (PVC). Engineering plastics exhibit superior mechanical and thermal properties, and they are represented by polyamides (PA), polycarbonates (PC), polyphenylene oxide (PPO), polyethylene terephthalate (PET) and polybutylene terephthalate (PBT). Table A5 shows the list of the representative polymeric materials. Polymeric materials are strengthened with fibers and hard particles to form composites, and the materials are hardened and endowed with excellent heat resistance.

A3.4. Semiconductors

Although pure silicon is basically an electric insulator, small amounts of doped elements such as phosphorus and boron increase the conductivity. The doping of P to Si results in n–type semiconductor because five M–shell electrons are present in P while Si has only four M–shell electrons. In the crystal structure of silicon, four electrons of each atom contribute to covalent bonding with no contribution to the electrical conductivity. The excess electrons contribute to electrical conductance, and phosphorus provides the excess electrons, resulting in n–type semiconductors.

Similarly, the doping of B to Si results in p–type semiconductor. Boron has only three M–shell electrons, and this is too small number to fill the orbital for the covalent bonding. The vacancies in the covalent bonding behave similar to the bubbles in water. The electrons forming covalent bonding are no more constricted, and move to the vacant sites. Under the electric field, the vacant sites look like moving toward the direction opposite to the movements of electrons. This behavior can be recognized as the vacant sites have positive charge, and the movement of the vacant sites may be expressed as the hole having positive charge. This behavior is similar to the air bubble movement in water, and the electron mobility is in general greater than the mobility of holes.

Semiconductors are used for diodes and transistors, and integrated circuits are useful for signal processing.

Table A5. Representative polymeric materials

Common plastics	LDPE	Low–density polyethylene
	HDPE	High–density polyethylene
	PP	Polypropylene
	PS	Polystyrene
	PVC	Polyvinyl chloride
Engineering plastics	PA	Polyamides
	PC	Polycarbonates
	PPO	Polyphenylene oxide
	PET	Polyethylene terephthalate
	PBT	Polybutylene terephthalate

A4. PHYSICAL PROPERTIES

Mechanical and thermal properties are associated with the strength of chemical bonds, and the states of electron are related to electric, magnetic and optical properties

A4.1. Mechanical Properties

The distance between neighboring atoms is elongated by mechanical forces, and elastic deformation occurs when the atoms are returned to the original position after the removal of the forces. When the forces are greater than the elastic limit, plastic deformation occurs due to the sliding of atomic planes. During the plastic deformation, atomic planes slide one another to form eternal deformation after the removal of the load. However, in the case that the load is too high, the separation of atomic planes occurs, and the crack is generated to propagate to final failure.

Elastic Modulus

Elastic modulus is defined as a ratio of stress to strain in linear elastic deformation. Young's modulus E is elastic response to tensile stress, and the shear modulus G is the response to shear deformation. Assume that the rod of length L with cross section area A_r is subjected to the tensile stress, as shown in Figure A5, and the rod is elongated with a decrease in the cross section. The Young's modulus E may be defined from strains and stresses as follows

$$E = \frac{\sigma}{\varepsilon} \tag{A.1}$$

where σ is the tensile stress and ε is the tensile strain

The similar relation may be defined in shear loading, as shown in Figure A6, as

$$G = \frac{\tau}{\gamma} \tag{A.2}$$

where τ is the shear stress and γ is the shear strain

These values are associated each other in isotropic materials as

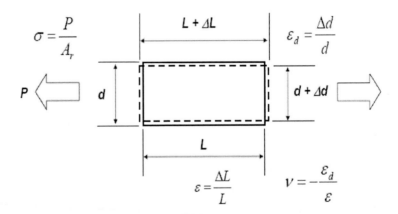

$$\sigma = \frac{P}{A_r} \qquad \varepsilon_d = \frac{\Delta d}{d}$$

$$\varepsilon = \frac{\Delta L}{L} \qquad \nu = -\frac{\varepsilon_d}{\varepsilon}$$

Figure A5. Elastic deformation under the tensile stress, the stress σ is defined by P/A_r, where P is a tensile load and A_r is the cross section area. The tensile strain ε is defined as $\Delta L/L$, and the length L is elongated to the length of $L + \Delta L$ under the tensile stress. The width perpendicular to the loading axis is shortened from d to $d + \Delta d$, and the negative strain ε_d is defined as $\Delta d/d$. The Poisson's ratio ν is defined as $\nu = -\varepsilon_d/\varepsilon$.

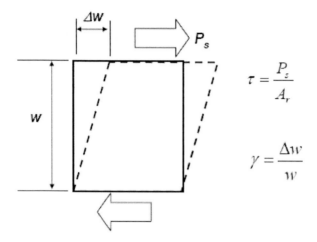

Figure A6. Elastic deformation under the shear stress, the stress σ is defined by P_s/A_r, where P_s is a shear load and A_r is the cross section area. The shear strain is defined as γ = $\Delta w/w$, and Δw is a displacement under the shear stress.

$$G = \frac{E}{2(1-v^2)} \qquad (A.3)$$

where v is Poison's ratio, denoting the ratio of perpendicular strain to tensile strain under the tensile stress.

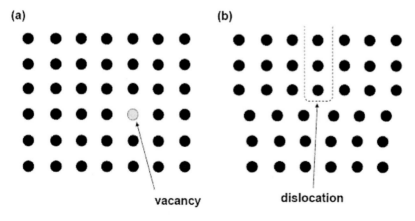

Figure A7. Imperfections in solids, (a) vacancy as a lattice defect, (b) dislocation as a line defect.

Deformation and Hardness

Under tensile loading, materials are elastically deformed up to some extents, and the yield stress is a limit of stress to initiate the plastic deformation. Plastic deformation intensively occurs at higher stress than that of yield stress. In general, the plastic deformation in metals occurs with slip of crystal planes under shear stresses, and the presence of dislocations in the crystals greatly reduces the stress required for deformation.

Although most of the atoms in crystals are regularly arranged, the crystallographic imperfections such as point defects and line defects are inevitably present, as shown in Figure A7. The dislocation is a line defect present in a crystal, and misalignment of atomic planes reduces the stress required for shear deformation in comparison with ideal crystals with no defects due to the partial misalignment present in atomic planes. The partial misalignment enables the localized movement of atomic planes during the shear deformation, and this greatly reduces the stress required for shear deformation since the ideal crystals require the very high stress for sliding whole planes.

However, the high density of dislocation makes further plastic deformation difficult because the movement of a dislocation is interfered each other. This is a mechanism for work hardening that metals become hard after experiencing a large deformation

Dislocation movement is enhanced at elevated temperatures due to the thermally activated atomic movements. In addition, the time–dependent deformation is enhanced at elevated temperatures due to the activated diffusion to cause the mass transportation. Creep is a time–dependent deformation occurring at elevated temperatures, and the materials are gradually deformed under the stress. The dislocation movements and lattice diffusion increase the creep rate.

Hardness is a confusing concept indicating the resistance to plastic deformation. The Vicker's hardness (H_v) is defined as the load per concaving area, and the similar definition of hardness is used for Knoop hardness (H_k) and Brinell hardness (H_B). The Rockwell hardness is defined as the concave depth, and Shore hardness is a repulsive hardness defined as the jumped height of hammer dropped on the specimen surface. Hardness is generally high for brittle materials and low at elevated temperatures.

Strength and Fatigue

The separation of crystal planes produces cracks, and the cracks propagate to cause the final failure. The yield stress is usually lower than fracture strength in metallic materials while fracture may occur in brittle materials such as ceramics under lower stress than that yielding occurs.

Fatigue is a phenomenon of the brittle failure after experiencing cyclic loading, and the fatigue failure occurs without exhibiting considerable deformation prior to final failure in cyclic loading of metals.

Fracture Toughness

The final fracture occurs as a result of crack propagation, and the fracture is sensitive to the existence of small cracks in brittle materials. On the contrary, ductile materials are insensitive to the crack, and the fatal fracture occurs after developing the crack to the large dimension.

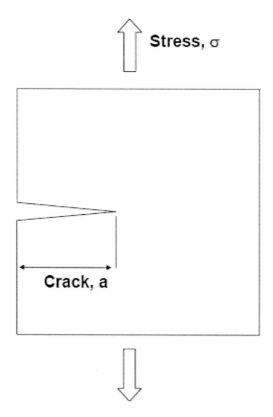

Figure A8. Configuration of a crack under the tensile stress.

The fracture sensitivity to the crack dimension is evaluated with fracture toughness. Under the same stress levels, large cracks are allowed in the materials having high fracture toughness, and the low fracture toughness leads to the final failure from a small crack.

The fracture toughness K_{IC} is defined by

$$K_{IC} = Y\sigma_T\sqrt{a} \tag{A.4}$$

where σ_T is tensile stress, a is a dimension of the crack, and Y is a geometrical factor depending on the shape of the crack and loading configuration. Figure A8 shows the configuration of the crack.

A4.2. Thermal Properties

Material properties are temperature dependent, and the materials change their form dependent on temperatures. Phase transition is a structural change in a material, and a solid is melt and vaporized at elevated temperatures. In addition, the materials may change the crystal structures dependent on the temperatures.

The averaged distance between neighboring atoms is elongated with a rise in temperatures. The materials melt at the temperature, where the lattice vibration becomes severe and the original positions of constitutional atoms are difficult to maintain. Vaporization occurs at higher temperature.

Specific Heat

Specific heat C_p is thermal energy required for a unit mass to rise the temperature by one degree under constant pressure. Since the temperature of a solid basically corresponds to the kinetic energy of vibrating constitutional atoms, the molar heat capacity at constant volume has been thought constant. The Dulong–Petit law indicates that molar heat capacity of solid is $3R$, where R is gas constant. This is a good approximation except for low temperatures.

The specific heat reaches to zero at 0 K, and this behavior was first explained by the Einstein model, and the model was further corrected with the advanced Debye model. Figure A9 shows the calculated dependence on temperature, and two models clearly show that specific heat steeply decreases in low temperature.

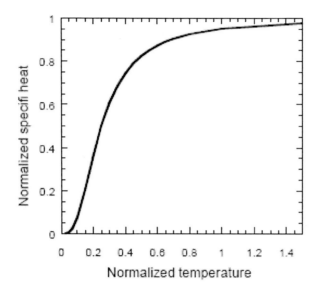

Figure A9. Specific heat as a function of temperature, predicted by the Debye model.

Thermal Diffusivity and Thermal Conductivity

Thermal diffusivity is a measure of adjustability in the temperature of a substance, and the substance of high thermal diffusivity rapidly adjusts their temperature to that of their surroundings. Thermal conductivity is a measure for ability of transmitting heat, and the materials of high thermal conductivity transfer the great amount of thermal energy. Thermal conductivity k and thermal diffusivity α are associated each other.

$$k = \alpha \, C_p \, \rho \tag{A.5}$$

where ρ is the density of the material, and C_p is specific heat at constant pressure.

Thermal conductivity is associated with electrical conduction in metals and lattice vibration in insulators. In metals, the ratio of the electronic contribution to the thermal conductivity and the electrical conductivity σ is proportional to the temperature, as indicated by the Wiedemann–Franz law

$$k / \sigma = L \, T \tag{A.6}$$

where T is absolute temperature and L is Lorenz number, and the Lorenz number is theoretically calculated as $2.45 \times 10^{-8} \, W\Omega/K^2$. In insulators, lattice vibration is responsible for thermal conduction, and the lattice vibration behaves as waves as well as particles. A phonon is a quantized mode of vibration occurring in a rigid crystal lattice, and the thermal conductivity is reduced with scattering of phonons. Therefore, the presence of impurities and imperfection in crystals reduces the thermal conductivity, and intensive lattice vibration occurring at high temperature scatters the phonons to lower the thermal conductivity in insulators.

A4.3. Properties and Chemical Bonding

Strength of chemical bonding governs the material properties, such as elastic modulus, thermal expansion coefficient, and melting temperature. Figure A10 shows the bonding energy as a function of the distance between atoms. The atomic position is normally in the bottom of the bonding energy curve, and the strong bonding force corresponds to the deep curve at the atomic position.

However, the atomic positions in the crystals fluctuate due to lattice vibration, and the distance between atoms is extended with an increase in the temperature. Finally, the materials melt above the melting temperature. Mechanical forces also shift the position, and the dimensional change in the distance between atoms is proportional to the applied stress and the coefficient determines the elastic modulus. Therefore, strong chemical bonding results in low thermal coefficient, high melting temperature, and high elastic modulus.

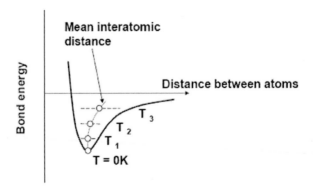

Figure A10. Bonding energy as a function of the distance between atoms.

Figure A11 shows the relation between Young's modulus and melting temperature for several materials. In general, materials of high Young's modulus exhibit high melting temperatures. Figure A12 shows the relation between thermal expansion coefficients and melting temperatures for metals, substances having diamond structures (Sn, Ge, Si and C), alkali halides and metal oxides. Except for the diamond structures, the materials of low melting temperatures have in general large thermal expansion coefficients.

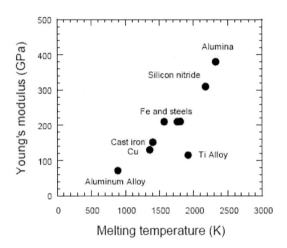

Figure A11. Young's modulus and melting temperature for representative materials.

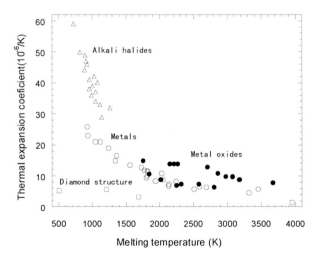

Figure A12. Thermal expansion coefficient and melting temperature for metals, diamond structures (Sn, Ge, Si and C), alkali halides and metal oxides (Original data are presented in: T. Nakamura, *Ceramics and heat*, Gihodo, Tokyo, 1985).

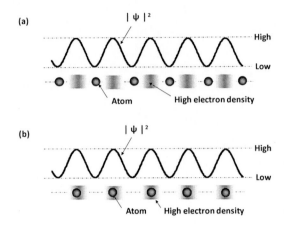

Figure A13. Two states of wave function ψ suggesting different energy states, (a) electrons are present between atoms, (b) electron are present very closed to the atoms. Note that the density of possibility is given by $|\psi|^2$.

A4.4. Electric Properties

The energy band model is used to describe the electronic state of solids. In the model, the atomic nuclei are fixed on the crystal lattice at regular positions, and the electrons in the inner shells of each atom are tightly bound to the atom. However, electrons in the outer shells are not bound to any particular atoms and free to move around in the crystal lattice. Therefore, free electrons are moving over the regularly arranged positive ions.

Quantum Theory

According to the quantum theory, electrons behave as particles as well as waves. In general, the wave may be expressed as a function of position x and time t.

$$y = A\sin(kx - \omega t + \alpha) \tag{A.7a}$$

where k, ω and α are the wave number , the angular velocity of the wave and the phase at $x = t = 0$, respectively. Equation (A.7a) may be reduced to

$$\frac{\partial^2 y}{\partial x^2} + k^2 y = 0 \tag{A.7b}$$

Schrödinger wave equation is derived from the law of the conservation of energy, and the kinetic energy is expressed in a form of wave characteristics assuming the de Broglie wave proposed in 1924, in which the units of energy E and momentum p are given by hv and h/λ, respectively. Here, h denotes the Planck constant, and λ and v are the wavelength and the frequency, respectively. It should be noted that $\omega = 2\pi v$ and $k = 2\pi/\lambda$.

Assume the total energy E is the sum of potential energy V and kinetic energy T, where $T = p^2/2m$. Accordingly, the law of the conservation of energy coupled with Equation (A.7b) leads to the time-independent Schrödinger equation.

$$\frac{\partial^2 \psi}{\partial x^2} + \frac{8\pi^2 m}{h^2}[E - V(x)]\psi = 0$$

which may be expressed in the three-dimesional form

$$E\psi = -\frac{h^2}{8\pi^2 m}\left\{\frac{\partial^2 \psi}{\partial x^2} + \frac{\partial^2 \psi}{\partial y^2} + \frac{\partial^2 \psi}{\partial z^2}\right\} + V(x, y, z)\psi \qquad (A.8)$$

Note that wave function of the Schrödinger equation is not a usual wave but indicates the density of possibility by $|\psi|^2$, indicated by M. Born in 1926.

Energy Band Models

Assume an electron in a unidirectional box with a length of L, and the magnitude of wave must be zero at the ends of the box, where $x = 0$ and $x = L$.

$$\psi = A\sin\frac{n\pi}{L}x \qquad (A.9)$$

where n is a positive integer. In the case of the potential energy V of zero, the momentum p calculated from n and L, is directly connected to the total energy E.

However, the potential energy is changed due to the existence of the regularly arranged positive ions. The total energy of electron is roughly proportional to the square of the momentum except for the particular wavelength as shown in Figure A13. This is an origin of the energy band gap, where no energy levels exist.

Electrons occupy the energy states depending on the temperature. Usually, most of electrons are in the lowest states, and some of electrons are activated in high energy levels. Accordingly, most of electrons fill the low levels in the energy band. In metals, high electrical conductivity is due to the partially filled band, and insulator materials have filled bands as shown in Figure A14. In semiconductors, small numbers of charge carrier such as electrons and holes are activated

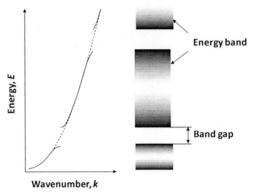

Fig. A14

Figure A14. Energy band model, indicating possible energy states for electrons. Electrons are not allowed to have energy between the energy bands, and the inhibited energy levels are stated as band gap.

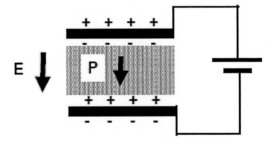

Figure A15. Polarization in dielectric materials.

Electrical Conductivity

Electrical conductivity σ (Ω^{-1} m^{-1}) may be expressed in terms of the carrier density n (m^{-3}), mobility or velocity of carrier μ (m^2 V^{-1} s^{-1}), and the electric charge of carrier q (C):

$$\sigma = q\,n\,\mu \tag{A.10}$$

In metals, the conductivity at high temperatures is reduced because the free electrons are frequently scattered. This reduces the mean free path of electron, leading to the reduction in mobility. However, thermally activated carriers in semiconductors increase at high temperatures, and the increased carrier concentration enhances the electrical conductivity.

A4.5. Dielectric Properties

As shown in Figure A15, the polarization P of materials under the electric field of E may be described as

$$P = \chi_e \varepsilon_0 E \tag{A.11}$$

where χ_e is the electric susceptibility, and ε_0 is the electric constant (or vacuum permittivity). The electric displacement field D may be given by

$$D = \varepsilon E \tag{A.12a}$$

or

$$D = \varepsilon_0 E + P \tag{A.12b}$$

where ε is the frequency-dependent absolute permittivity of the material. The dielectric constant (or relative permittivity of the material) ε_r may be defined as

$$\varepsilon_r = \varepsilon_0 / \tag{A.13}$$

Dielectric Materials

In insulator materials, electrons of each atom are tightly bound to the atom. However, an electric cloud of negative charge can be changed its shape when electric field is applied. The materials are polarized due to the distortion of the electric cloud, as well as the distortion of dipoles and ions.

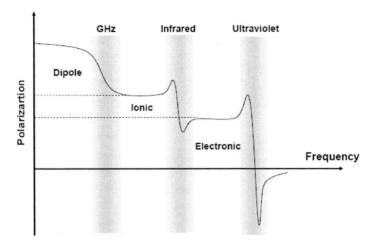

Figure A16. Changes in polarization as a function of frequency. The contribution of dipole, ionic and electronic polarization depends on frequency.

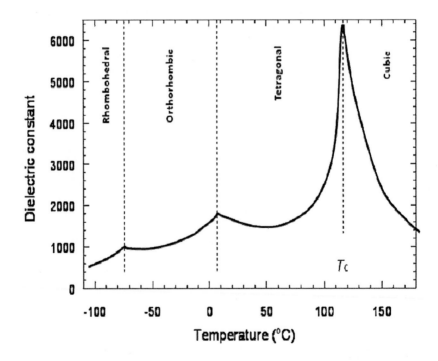

Figure A17. Dielectric constant of ferroelectric $BaTiO_3$ as a function of temperature. Crystal structure changes with temperature, and the Curie temperature is at the transition between tetragonal and cubic crystal structures.

Since polarization occurs due to the movement of charged particles, the dielectric constant depends on the frequency of electric field. Figure A16 shows that dipole distortion occurs at relatively low frequency up to the microwave frequency, and ionic distortion occurs up to the frequency of infrared light. In addition, electronic distortion occurs even at higher frequency of visual light. Although the polarization greatly changes at transitions in distortion mechanisms, the dielectric constant is high in low frequency and low in high frequency. The dielectric constant is in general higher than unity but not so high except for ferroelectric materials.

Ferroelectricity

Some classes of materials exhibit very high dielectric constants due to large distortion. Ions in ferroelectric materials are shifted to another semi–stable site when the external electric field is applied. In ferroelectric materials in the perovskite structures of ABO_3, there are two stable positions for B–site ions in the crystal structure, and the B–site ions are movable between two possible positions. This causes large polarization accompanying a hysteresis loop in the $P–E$ curve. However, the actual values of dielectric constants become large when the crystal structure is unstabilized. The capacitance of ceramic capacitor, represented by dielectric material of $BaTiO_3$, is actually high when the temperature is closed to the Curie temperature of the ferroelectric materials (Figure A17). At temperatures closed to the Curie temperature, the several stable positions of B–site ions Ti are created in the crystal structure.

Piezoelectricity

Piezoelectricity is the ability of materials generating electric field as a response to applied mechanical forces, and is related to the unsymmetrical configuration of crystal structure. Various materials, including quartz and bone, exhibit piezoelectricity while the poled ferroelectric materials exhibit the large effect in piezoelectricity. The representative material is $Pb(Zr, Ti)O_3$, which is denoted as PZT and situated around midst in a solid solution of $PbZrO_3$ and $PbTiO_3$. The large piezoelectric deformation is thought to result from the morphotropic phase boundary, where is close to the coexistence composition of tetragonal to rhombohedral phases.

Piezoelectricity of quartz is widely applied to generating standard frequency while the piezoelectric strain is small in comparison with piezoelectric ceramics such as PZT. Piezoelectric material vibrates at a regular frequency depending on the dimension when a periodic electric field is applied to the crystal. Quartz crystals cut with a particular crystal plane exhibit a very stable resonant frequency of vibration nearly independent of temperature, and applied to the standard frequency of electric clocks.

A4.6. Magnetic Properties

The magnetization M of the material under the magnetic field strength H may be described as

$$M = \chi_m H \tag{A.14}$$

where χ_m is the magnetic susceptibility. The magnetic flux density B may be expressed as

$$B = \mu H \tag{A.15a}$$

or
$$B = \mu_0(H + M) \tag{A.15b}$$

where μ is the permeability of the material and μ_0 is the magnetic constant (or vacuum permeability). The relative permeability μ_r may be defined as

$$\mu_r = \mu/\mu_0 \tag{A.16}$$

Diamagnetism
The magnetic susceptibility is derived from magnetic response to the applied magnetic field, and the diamagnetism is the property of materials having negative magnetic susceptibility. In diamagnetic materials, the electrons present in the orbital are coupled.

Paramagnetism
The positive susceptibility is due to the response of magnetic moment of the material. Since the rotating electric current produces the magnetic field, an uncoupled electron in an orbital gives the magnetic moment. In the case that a couple of electrons are present in a single orbital, the magnetic moments are cancelled each other and contribute no positive susceptibility. However, positive susceptibility is found in metallic materials and materials having partially filled orbital, where a single electron is present in a single orbital.

Paramagnetism is the property of materials having small positive magnetic susceptibility.

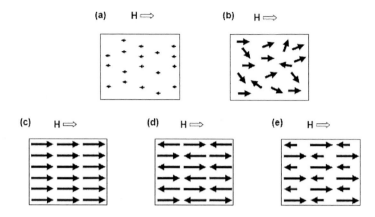

Figure A18. A variety of magnetisms under applied magnetic filed, (a) diamagnetism, (b) paramagnetism, (c) ferromagnetism, (d) antiferromagnetism, (e) ferrimagnetism.

Magnetic Moment

In general, the magnetic susceptibility is very small and the effect of magnetization of materials on the flux density is negligible. However, some materials exhibit large magnetic susceptibility because of magnetization. Magnetization occurs by aligning the directions of magnetic moments in the material as a response to the applied magnetic field, and strong interaction between the magnetic moments may order the direction of the moments. The ordering of magnetic moments may occur in two ways: one is to order in the same direction and the other is to order in the opposite direction between neighboring atoms. Figure A18 shows a variety of magnetisms in a view of alignments in the magnetic moment.

Ferromagnetism and Ferrimagnetism

The magnetic moments are ordered in same direction in ferromagentic materials, such as iron, which exhibit very large permeability. Antiferromagnetism is a result of ordering in the opposite direction, and the total magnetic moment is completely cancelled, and magnetic susceptibility is negligible. On the contrary, ferrimagnetism is in a similar interaction to antiferromagnetism between the magnetic moments but the magnitude of the magnetic moments between the neighboring atoms is different. Therefore, the total of the magnetic moments produces the large permeability in ferrimagnetic materials represented in ferrite ceramics.

Permanent Magnets and Domain

A large permeability in ferromagnetic and ferrimagnetic materials is associated with domain structure. The magnetized regions are stabilized when small areas of opposite magnetizing directions are coupled under low magnetic field, and the domain is the region of the area as shown in Figure A19. If the domain structure is promptly changed with the applied magnetic field, B–H curves are typically linear and no hysteresis is seen. The classes of materials are useful for magnetic core and applied to electromagnets. However, large hysteresis is realized for the materials, in which obstacles present in domain walls interrupt the change in the domain structure. In this case, the material is a permanent magnet because the residual magnetization is present even after the removal of applied magnetic fields.

A4.7. Optical Properties

Transparency is the physical property of allowing light to pass through a material, and the incident light is partially refracted and reflected when the light passes from one medium to another.

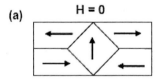

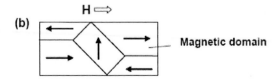

Figure A19. Changes of magnetic domain structure, (a) no magnetic field, (b) applied magnetic field

The refraction of light is associated with the refractive indices, which are given by the ratio of light velocities between two media. The intensity of light beam is reduced during traveling in the medium due to absorption and scattering. The reflection intensively occurs in metallic materials where the numerous free electrons exist.

Refraction

According to the Snell's law, the angle of incidence θ_1 is related to the angle of refraction θ_2 by

$$\frac{n_2}{n_1} = \frac{v_1}{v_2} = \frac{\sin \theta_1}{\sin \theta_2} \tag{A.17}$$

where v_1 and v_2 are the wave velocities in the respective media, and n_1 and n_2 the refractive indices, as shown in Figure A20. The refractive index is therefore a measure of how much the speed of light is reduced inside the materials.

Note that the refractive index of a material is associated with the dielectric constant and relative permeability

$$n = \sqrt{\varepsilon_r \mu_r} \tag{A.18}$$

In most of materials, the relative permeability of the material is close to unity, hence

$$n = \sqrt{\varepsilon_r} \tag{A.18a}$$

The change in the direction of light is associated with the refractive indices, which are given by the ratio of light velocities between two media. It should be noted that the refractive index depends on the frequency of light, and the different colors of the incident light to the medium take different passes. This is the reason why the glass prism produces the seven different colors under sunshine. In general, the refractive index is high for blue light and low for red light. This is because the frequency of blue light is close to the ultraviolet light, in which resonance absorption occurs in condensed matter to result in large polarization, as shown in Figure A16. The large polarization due to the resonance absorption in ultraviolet frequency causes high dielectric constant leading to high refractive index. Accordingly, the refractive index is higher at the higher frequency of the visual light region.

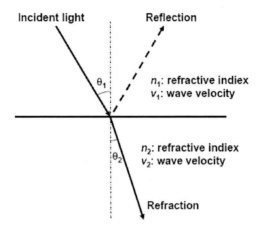

Figure A20. Optical reflection and refraction indicating that the incident light travels from medium 1 into medium 2, and a part of light beam is reflected at the interface.

Reflection

Reflection is the change in direction of a wavefront at an interface between two different media so that the wavefront returns into the medium from which it originated. Optical reflection occurs in metallic materials, where free electrons are present. Electromagnetic wave forces free electrons under alternating electric field, and this inhibits the propagating of light in metallic materials.

Plasma Frequency

In the case that the frequency of electromagnetic wave is so high, light can propagate into the materials having free electrons. This indicates that electromagnetic waves of low frequency are unable to propagate in the metallic materials but high frequency waves are able to propagate in the metals. Plasma frequency is the transition between them. Most of metals have a plasma frequency in ultraviolet region while some materials such as cupper and gold seem to have the plasma frequency in the visible light region. The plasma frequency is associated with charge density, and semiconductors have a plasma frequency in infrared regions represented with transparent semiconducting oxides such as indium tin oxide (ITO).

Absorption

Light has dual wave–particle nature of electromagnetic waves and photons. The photon is a quantum of the electromagnetic field, and is absorbed when the energy of the incident photon is consumed for exciting the quantum state of an electron and an atom. Light is intensively absorbed when the energy of photon is equal to the gap in quantum energy levels. Transition in electron levels is generally in the region of visible and ultraviolet light, and the strong absorption of light normally corresponds to the transition between energy levels equal to or greater than the band gap. The transition in vibration energy states is in the region of infrared frequency, and the energy levels of the rotational motion are much low in microwave frequency.

The presence of absorbing sites such as impurities and defects in the materials is responsible for the optical absorption. Elemental impurities in the materials absorb and emit the light of wavelength in accordance with the electronic structures. The difference in energy levels between the ground and excited states corresponds to the wavelength of absorbing and emitting light.

Luminescence is the visible light emission from a substance where electrons have been transferred to the excited states, and stimulation such as light of shorter wavelength, a strong electric field, and an electron beam radiation is able to produce the activated state.

APPENDIX B: PHYSICAL CONSTANTS

Important physical constants are listed in Table B1.

Table B1. Important physical constants*

Quantity	Symbol	Values	Note
Speed of light in vacuum	c	$2.99792458 \times 10^8 \, \text{m·s}^{-1}$	
Elementary charge	e	$1.602176487 \times 10^{-19} \, \text{C}$	
Planck constant	h	$6.62606896 \times 10^{-34} \, \text{J·s}$	
Boltzmann constant	k_B	$1.3806504 \times 10^{-23} \, \text{J·K}^{-1}$	
Gravitational constant	G	$6.67428 \times 10^{-11} \, \text{m}^3 \text{·kg}^{-1} \text{·s}^{-2}$	
Avogadro's number	N_{Av}	$6.02214179 \times 10^{23} \, \text{mol}^{-1}$	
Electric constant	ε_0	$8.854187817 \times 10^{-12} \, \text{C}^2 \, \text{N}^{-1} \, \text{m}^{-2}$	$\varepsilon_0 = 1/(\mu_0 c^2)$
Magnetic constant	μ_0	$4\pi \times 10^{-7} \, \text{N·A}^{-2}$	Defined
Gas constant	R	$8.314472 \, \text{J·K}^{-1} \text{·mol}^{-1}$	$R = N_{Av} \, k_B$

* Mohr, P. J., Taylor, B. N. and Newell, D. B., CODATA Recommended Values of the Fundamental Physical Constants: 2006, *Rev. Mod. Phys.*, 80 [2], 633-730 (2008).

INDEX

D